Why the Wheel Is Round

Why the Wheel Is Round

Muscles, Technology,
and How We Make Things Move

Steven Vogel

The University of Chicago Press
Chicago and London

The University of Chicago Press, Chicago 60637
The University of Chicago Press, Ltd., London
© 2016 by Steven Vogel
All rights reserved. Published 2016.
Printed in the United States of America

25 24 23 22 21 20 19 18 17   3 4 5

ISBN-13: 978-0-226-38103-9 (cloth)
ISBN-13: 978-0-226-38117-6 (e-book)
DOI: 10.7208/chicago/9780226381176.001.0001

Library of Congress Cataloging-in-Publication Data

Names: Vogel, Steven, 1940–2015, author.
Title: Why the wheel is round: muscles, technology, and how we make things
    move / Steven Vogel.
Description: Chicago: The University of Chicago Press, 2016. | Includes
    bibliographical references.
Identifiers: LCCN 2016005058 | ISBN 9780226381039 (cloth: alk. paper) |
    ISBN 9780226381176 (e-book)
Subjects: LCSH: Biomechanics. | Rotational motion.
Classification: LCC QH513 .V644 2016 | DDC 612.7/6—dc23 LC record
    available at http://lccn.loc.gov/2016005058

To Jane,
with love and appreciation,
now more than ever

[**Contents**]

[Preface]

As surely as we remain animals, biology establishes the baseline for
what we do; it did so even more pervasively in our past than it does
at present. Just as (to quote Lincoln) we cannot escape history, quite
as certainly, our history cannot escape biology. Of course that's easy
to say, but it's too multifaceted to expound upon with any brevity—
parasitology, population genetics, plant domestication—just to pick
relevant headings that begin with the letter p. One book could not
possibly do the subject justice.

Both by profession and mental habit, I remain a biologist, even if
the word "interdisciplinary" has become ever less appropriate than
"undisciplined" now that I have the intellectual freedom afforded by
formal retirement. Beyond that, I'm an unreconstructed academic,
with the academic's peculiar willingness to look elsewhere than the
applications of science and look instead at origins, underpinnings,
and interrelationships.

No great amount of sleuthing is needed to find out that my basic
field is biomechanics. Here that subject worries about how muscles,
by pulling on bones, allow us to do our ordinary tasks, plus how the
properties of biological materials such as wood, horn, shell, and the

like fit them for toolmaking. But I'm also indulging my long avocational interest in history, in particular the history of technology, plus a growing interest in anthropology—all seen through the peculiar lens (or maybe kaleidoscope) of a biomechanic. The upshot, though, mixes, in addition to those areas, helpings of archaeology, mechanical engineering, and physics, along with bits of cultural, political, and military history. I've been nothing if not indulgent in casting a wide net for items of at least arguable relevance.

A few words about sources. Of course the usual ones—books, journal articles, colleagues—contributed their usual generous service. Beyond those upon which I've relied on throughout my career, I now add the online ones. Almost every journal I use has now scanned its archives back to the year one, and a remarkable number of old technical works have become available through the kind offices of, especially, www.archive.org. We all use Google, admirable if one has a good search term, and better than most people realize if one takes advantage of its full capabilities to handle a complex search strategy. I'm especially fond of Google Scholar; not only does it focus on scholarly and technical literature, but, *mirabile dictu*, it permits forward searching by clicking on "cited by." Thus if one has some classic paper or book, you can find out who gives it as a reference and thus work your way up to the very present—the opposite of working backward through bibliographies. And I must say good things about *Wikipedia*, too often maligned in academic circles. I find it splendid—imperfect, of course, as are all other sources, as one therefore should expect, but far better than one expects or, perhaps, deserves. The one really misleading article that I might have cited here has been corrected (no, I didn't instigate the correction), and, yes, I do make an annual contribution.

One old resource has proven especially valuable, the more so with so much of it now accessible or at least searchable online. Anthropology took off during the nineteenth century, with a shift from collectors to trained observers visiting the cultures just then on the verge of the loss of identity and traditions concomitant with modern communications. These anthropologists carried few if any cameras in that era before roll film replaced glass plates, so they could depend only on drawings to supplement their words. Those drawings, skillfully and sometimes even artistically done, remain as a record, entirely in the public domain.

Each time I start a book, I promise myself that I will faithfully keep a list of the people who have helped in its creation—providing me with all manner of information, correcting my misconceptions, informing me of significant sources, reading preliminary prose, and so forth. And each time I don't do the job as well as I ought to. So with that disclaimer to recognize the incompleteness of the list, I must thank David Arons, Kalman Bland, Caroline Bruzelius, Steve Churchill, Ed Dougherty, Donald Fluke, Henry Halboth, Bob Healy, Charlie Henderson, Maggie Hivnor-LaBarbera, Michael LaBarbera, Sy Mauskopf, Chuck Pell, Steve and Kathy Rostand, Gillian Suss, Jane Vogel, and Bob Wallace. I'm particularly indebted to Christie Henry of the University of Chicago Press for dealing with some unusual aspects of preparing this book. Plus I thank two very helpful anonymous reviewers.

Various groups have been semi-willing test audiences for the material of the present book. Among these are some fourth-grade students at Livesey Elementary School in Tucker, Georgia (instigated by Avery Vogel, an F2); some fourth- and fifth-grade students at Club Boulevard Magnet Elementary School, Durham, North Carolina (at the invitation of Gary Krieger); and a group at the North Carolina Children's Hospital in the Healing and Hope Through Science Program of the North Carolina Botanical Garden (arranged by Katie Stoudemire and Tami Atkins). An early draft of the book was inflicted on a class in the Osher Lifelong Learning Institute at Duke University, given physically at Croasdaile Village, where I live and write.

[1]

Circling Bodies

First—don't be shy—try a few motions with your own body. Twist an extended arm as far as you can one way and then twist it the other way. Your wrist (mainly) can't even do a full 360-degree rotation. Twist your neck—your head won't rotate even as far as your hand did. Your lower back's mobility limits how far your torso can rotate just as severely, and feet (mainly ankles) feel still greater rotational constraint. All sorts of limbering and muscle-strengthening exercises depend on rotation—"curls" just put the matter more explicitly. After all, appendages hook on to us at pivot points around which they swing. But they swing through limited arcs, with varying degrees of constraint. Thus arms move around shoulders more freely; legs around hips less so, with flexibility evidently traded against stability and reliability. No picture need be provided; doing it yourself should be persuasive.

Continuous rotation, as with a proper wheel? For better or worse, no animal joint has ever managed that trick. Yes, we humans can rotate continuously—but only if we do it as a whole-body activity—as do somersaulting or rolling children. Almost all other creatures that rotate live within that general limitation as well. We're looking at

1

tumbleweeds, a shrimp that rolls back to the water when washed up on a beach,[1] a caterpillar that rotates head over heels, so to speak,[2] and the helicopter-like seeds (really fruits, technically samaras) of trees such as maples. More about these systems in a few pages.

Then look around. Sure, we've created a host of devices that may turn but also face (by design) much the same limitation on rotation—most hinges, door handles, light switches, latches, staplers, scissors, pliers . . . But playing a far more central role in our technology are things that rotate without limit as parts of otherwise non-turners, things that go around and around as long as they're driven and perhaps a little longer. I mean devices based on that marvelous invention, the wheel and axle. That includes almost all of our motors and their associated shafts, pulleys, gears, and so forth. It includes our diverse wheeled and propeller-driven vehicles. Plus all manner of hand tools, from eggbeaters to socket wrenches. Long ago that meant wagons and potter's wheels, and the diversity of our rotational contraptions has been on the increase throughout our history. No doubt at all—mechanisms that rotate as parts of otherwise non-rotating contrivances form the very core of our mechanical technology.

We thus glimpse a paradoxical problem. Through most of human history (and prehistory, if you prefer the distinction), muscle has been the main motor of our technology, whether we work our own personal meat or persuade that of our domesticated animals to do our jobs. Muscle can only pull, and it must remain attached at both its ends. How can a non-rotating engine drive truly rotational machinery? This book explores the diverse ways that humans have faced up to and managed to deal with that most basic of dilemmas. In essence, it explores one facet of the biomechanics behind history.

Your immediate rejoinder might be that the difficulty yields to a trivially simple fix. Specifically, just add a crank, a lever extending radially outward from the rotating shaft with a slip fitting on a sideways extension of that lever. No need for an illustration—we make such things all the time, from hand-operated household gear such as pencil sharpeners, eggbeaters, and meat grinders to the engines of our cars, in which pistons moving (for most cars) up and down crank and thereby turn driveshafts. That slip fitting might be nothing more than a greasy hand or a loosely fitting outer handle of wood or plastic. It seems reasonable that this obvious trick should have been particularly appropriate for ancient devices, with their slow rotation rates. Oddly enough, cranks remained unknown (or nearly so) until about a thousand years ago. Think of it—for all their sophistication, the classical Mediterranean civilizations made no significant use of this simple and now ubiquitous arrangement. Punning subtly, one might ask, where's the rub?

Muscle-powered rotational machinery obviously has a much longer history than cranks—think again about all those wagons, chariots, and potter's wheels. How, then, were they persuaded to rotate? And have these more ancient fixes persisted, even gained in importance, with the further proliferation of rotational devices? No surprise—one question leads to another.

First, then, what are the options for making shafts and wheels turn? If nothing else, its peculiar modernity tells us that a crank isn't the only thing that will work. Consider some other possibilities, put as a series that I don't assert is chronological, fully complete, or mutually exclusive—and at the expense of suspense . . .

- Roll the top of a cylinder by pushing something across it while the bottom then rolls (at half the speed) along the ground— rolling a log or barrel, as in figure 1.1. Of course, sooner or later (more likely sooner), the propelling roller on top leaves the driven roller behind. So you can't cover much distance without fairly often moving the roller left behind from rear to front. Even with a series of driven rollers, creating a new front one with a rear reject remains required. The simpler French-style or rod-type rolling pin works this way; its task doesn't ask that it roll very far and allows easy lifting and repositioning
- Pull or push on the axis of a wheel while a part of its circumference contacts the ground with enough friction so it

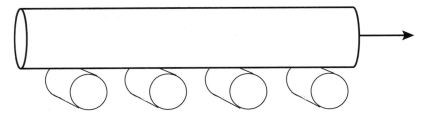

Figure 1.1. Moving something with a set of rollers beneath; obviously, it can't roll very far unless the rollers left behind are repositioned in front.

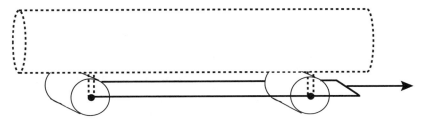

Figure 1.2. Moving something by pulling on the axis of a rotating wheel or set of wheels. This scheme solves the problem shown in the previous figure, but you then have problems of bearings and of attaching some carrier to the wheel(s).

rotates rather than just sliding along—as a horse pulls a cart and as in figure 1.2; or as you use a conventional rolling pin, one with a rotating handle at each end, by pushing or pulling the handles. The rolling pin then rotates as it presses the pie crust, although the handles do not. Proper bearings aren't absolutely necessary—a person can pull along a bagel-shaped (toroidal) water tank, hauling on a rope that loops through its center hole.

- Make an animal (perhaps a person) walk while pushing or pulling in monotonous circles around a vertical shaft or drum from which a radial lever protrudes—for example, turning a large posthole digger (auger), as in figure 1.3. The motor itself then rotates at just the same speed as the shaft or drum, so

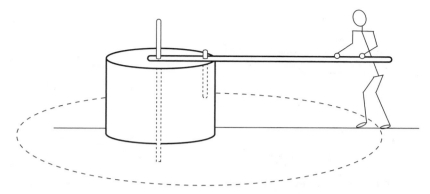

Figure 1.3. Here the entire motor, animal or person, moves in a properly rotational circle around a vertical shaft. Everything turns at the same angular rate. Of course, complexity ensues if one needs a non-vertical axis.

no bearing need be supplied—at least between the two. (Of course, that shaft or drum will typically turn around its own bearing.) Years ago, playgrounds had small merry-go-rounds driven by one or more children as others sat on the deck and made encouraging noises.

- Grab the handle of a tool, turn it through an arc, then release it, grab it again after turning one's arm or body some ways opposite the direction of the tool's rotation, and turn it again, as in figure 1.4. The prose may imply complication, but the process could not be simpler or more familiar. It's what we do with the steering wheels of cars and with the knobs on such electronic gear as still has knobs. And we do the same with screwdrivers and screw-on jar lids.

- Design the tool so the activity you're performing with it includes a recovery phase in which the tool's shaft rotates back to its original orientation—as with the knobs of old wristwatches and in figure 1.5. Thus no net rotation occurs, and no problem arises. A yo-yo works that way, as did many ancient tools— drills for boring holes and starting fires, for instance. Turn and re-turn, one might say.

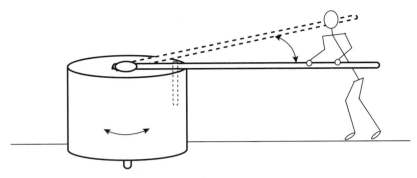

Figure 1.4. Turning through an arc presents no great problem, so many rotational devices work by turning an arc repeatedly, with some recovery phase between each turning episode—like turning a screwdriver. The arc indicated by the arrows represents the largest possible single arc of rotation.

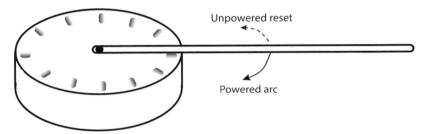

Figure 1.5. Here the tool incorporates the recovery phase, either with some ratcheting arrangement (a socket wrench) or else incorporating it as another power stroke, albeit turning in the other direction (boring with an awl). This circumvents the limited arc of the previous figure.

- Roll something, perhaps a rope or bundle of fibers, up on a shaft—a shaft with one end free from any supporting bearing. Every so often pull the roll off that free end of the shaft without unrolling it, as in figure 1.6. Each of the original rolling turns then becomes a twist of that rope or bundle of fibers. That's the basic trick behind spinning thread or cordage of any kind, in effect making long, tension-resisting, flexible material from the short fibers we harvest from plants (cotton, for instance) or

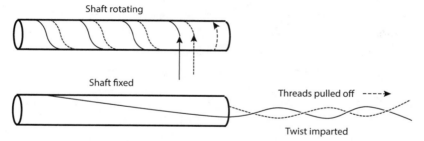

Figure 1.6. Rotation can be imparted by winding something flexible onto a rotating shaft and then pulling (perhaps periodically) that flexible something (usually a fiber bundle) off the free end of the shaft. As we'll see later, this is the basis of the oldest devices that spin thread or rope.

animals (wool and so forth). (Only the silk of silk moths comes in naturally long fibers, and we spin these mainly to bring them up to a convenient diameter for use as thread.)

We'll get back to each of these, exploring their advantages and disadvantages and how each has been used by technologies based on both muscle and other movers—in short, its functions, origins, and history. Of course, we have only spotty knowledge of the early history of devices as basic as these. Different cultures have taken different technological trajectories, and the extent to which the *how-to-do-its* of living have spread among them continues to generate controversy. Too often it's far from clear whether a technique was learned from another culture or whether it was independently invented—to say nothing of the matter of when either happened. Moreover, the history of technology has a problem of sourcing that's far worse than that of, say, the history of science. Craftsmen were not just secretive; until recently they were almost always illiterate. Science deliberately leaves a written record (even if it sometimes gets lost); technology rarely does so. Still, that may leave too bleak an impression—technology is the more likely of the two to leave behind some physical impression, a persistent archaeological record as artifacts. In addition, its history lives on in such things as common words and linguistic allusions. For instance, the expression "loose cannon" refers to the mayhem caused when a cannon of a sailing warship, weighing perhaps half a ton, came unhitched and careened around with the rolling of the ship, smashing almost anything in its erratic path.

Before going further, a distinction needs to be established, an absolutely critical matter here but one all too vague in everyday speech. Heading off in a constant direction will never be confused with rotation. But what if you move in a circle while never changing orientation, continuously facing the same direction? Admittedly this takes some unusual footwork inasmuch as, at times, you have to go sideways and backwards. If you trace your path on the floor, you certainly will find that you've made some kind of a closed loop, so you've undoubtedly gone around. At the same time, if you've faced the same way throughout, just as undoubtedly you haven't turned. So there are two ways to go around in circles. For practical reasons, mainly for describing motion with equations and for stating important conservation laws, physical scientists distinguish between these two kinds of circular motion. We need to do so as well.

Terminology. By definition, then, circular motion comes in two versions—not to exclude a mix of the two. In "rotation," orientation changes with time; in "translation," orientation doesn't change even if a body moves in a circle or part (an arc) of a circle. Figure 1.7 illustrates the difference. For present purposes, we'll rigorously restrict use of the term "rotation" to its proper physical kind. Yes, irrotational circular motion sounds oxymoronic, but clearly it's not. Moreover, it matters more than you might think. It takes on especial importance in fluid dynamics—as when a wing generates lift or a hurricane blows in a huge circle. We're really quite good at it ourselves, whether you exercise as a whole body or as you move a hand in a circle, signaling that someone might pass you. In American football, a ball carrier dodges and swerves and goes around while moving

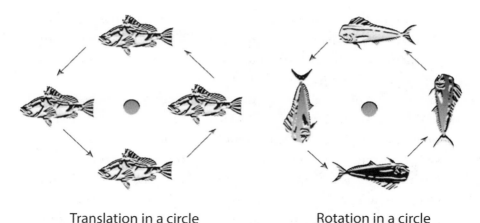

Translation in a circle   **Rotation in a circle**

Figure 1.7. Contrasting the two forms of circular motion. On the left, the fish translates around an axis like the cars on a Ferris wheel; on the right, the fish rotates around an axis like the frame of the Ferris wheel.

downfield, translating with the body ever facing the goal line. The carrier truly rotates only when shaking off a tackler with a whole body spin.

Our sensory equipment makes exactly this distinction, doing it without arousing your awareness. You translate in circles of any diameter and at any speed without getting dizzy, but when you rotate in circles, you have no such luck. Slow social dancing involves lots of circular translation, as do at least some maneuvers in square dancing. A Ferris wheel rotates, but its individual compartments, their orientation maintained gravitationally, translate in circles.[3] By contrast, ballet and ice dancing go in for vertiginous levels of true rotation, no trivial matter for the performer. Still, for even these last, the motions consist entirely of whole body rotation; again, that's the best we can do with our lack of fully rotational joints.

I would have preferred more descriptive designations emphasizing the contrast between, say, "motion with change of heading" for rotation and "motion without change of heading" for circular translation, but we're stuck with the oddly specialized use of two ordinary (and thus easily misunderstood) words. Early in the twentieth century, the psychologist and philosopher William James offered an excellent illustration, even if coupled with a message that we have to reject quite explicitly.[4] He imagined a hunter encountering a squirrel on the trunk of a tree. The squirrel runs around to the opposite side of the tree, so the hunter, at a much greater radius, moves around as well. The squirrel, no dodo, would like to survive the encounter, so it keeps moving in order to keep the trunk between itself and the hunter. Thus both squirrel and hunter make rotational motions. Does the hunter circle around the squirrel? He (male in the original) remains facing the squirrel, so he clearly does not. At the same time, he's north, then east, then south, and then west of the squirrel, so just as clearly he must circle the squirrel. James, illustrating the essence of pragmatism with the tale, said that the distinction is purely semantic and thus essentially meaningless. However, for our purposes, without a doubt both squirrel and hunter have engaged in true rotational motion, with the latter's motion describing a path around that of the former.

Not that translating around in a circle, without conversion into rotation by means of a crank, can't serve practical purposes. Think of what you do when stirring a pot or the batter for a cake. You make the stirring spoon translate around in circles, and it does its job at least as well as it would if it were truly rotating. A traditional mortar and pestle works the same way. These translational actions may even do better than their rotational equivalents—one translational turn will produce more movement of the pestle's periphery than would one rotational turn. Sometimes they can do very much better, since rotating a shaft in materials that retain odd traces of solidity often leads to undesirable effects—more on this business (strangulated flow) in chapter 10.

A complicated (and probably hypothetical) machine, a particularly ingenious contrivance, provides an especially neat and satisfying illustration of this distinction. Among much larger and more immediately important machines, Agostino Ramelli, a sixteenth-

Figure 1.8. Ramelli's reader, with his drawing of its epicyclic gearing.

century military engineer, designed a vertical wheel that kept a set of
books open for a single reader, as in figure 1.8.[5] By turning the wheel,
the end user (as we would now say) could select which volume to
consult, and volumes stayed both in a fixed orientation and opened
to preselected pages. So the wheel rotated but the individual book

supports (and books) translated in a circle. Ramelli accomplished the trick with what are called "epicyclic" or "planetary" gears; in this particular case, the central (sun) gear doesn't either rotate or translate, and the outer planetary gears translate but do not rotate. To effect this marriage, the planetary gears need to have the same number of teeth as the sun gears. (Neither the number of teeth on the intermediate gears nor the number of intermediate gears matters— they just ensure the correct relative direction of turning of the planets or, put strictly, assuring their non-turning.) Ramelli's is a particular (and odd) application of this kind of gearing, which was known if not common at the time. It appears in Leonardo da Vinci's notebooks, for instance, and it had been occasionally used in clock movements. We've used it in many automobile transmissions, from that of the Ford Model T to modern overdrives and automatics. A lovely animation of such epicyclic gearing appears in the *Wikipedia* article on gears.[6]

As well as introducing the underlying elements on which the story will turn, perhaps the author ought to expose his personal perspective. My main professional area has been biology, centering on biomechanics in the broadest sense—as might be suspected from an account that began with the range of motion of our appendages. As an experimentalist in an area without a stereotyped experimental armamentarium, I've repeatedly had to cobble together odd tools. I've long recognized that the more mechanical items one makes, the more adept one becomes at devising both quick fixes and generally useful pieces of apparatus. The various challenges, over more than fifty years, have often asked that I look into the state of one art or another—metalworking, devising simple electronic circuits, pipefitting, adapting motors, and so forth. Not only have I acquired some distinctly arcane abilities, but the problems, by yielding to solutions involving things no longer widely used, have often tickled my still older interest in history.

Back when I was a graduate student, I built a tiny anemometer to measure very low-speed airflow, one with a sensor well less than a millimeter in diameter. Calibration required flows of known speeds, and I had no reliable reference source. The easiest way that occurred to me to provide these minimal winds consisted of putting the sensor on the end of a beam that rotated around in the room at rates I could determine with a stopwatch. The device goes by the name "whirling arm," and it appears to have been invented about 1742 by the English scientist Benjamin Robins;[7] figure 1.9 shows his device. It was famously used by John Smeaton, the father of British civil engineering and builder of the great third Eddystone Lighthouse and many other structures. With a whirling arm (as well as with other clever equipment), Smeaton produced the first tables from which the performance of waterwheels and windmills could be calculated.[8] In Smeaton's version, more elaborate than that of Robins, two descending weights turned capstans, which in turn rotated the arm and an attached propeller. In fact, neither version achieves continuous rotational motion. Returning to our earlier list of options, rewinding reversed the rotation as it raised the weights, as in figure 1.5. (Nor did mine—turning twisted a loose bundle of wires, so it needed

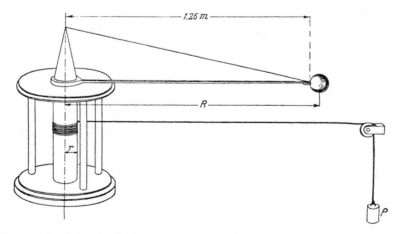

Figure 1.9. Robins's whirling arm apparatus for measuring drag. The weight (p) insures a constant torque so the speed of rotation is inversely proportional to the drag of the object.

periodic unturning.) And the whirling arm was almost as famously used by the German engineer and aviation pioneer Otto Lilienthal to generate tables of lift and drag as he worked toward flying machines late in the nineteenth century.[9]

As it turned out, data from both Smeaton and Lilienthal suffered from serious systematic errors, errors discovered (to their initial dismay) by Wilbur and Orville Wright in the early years of the twentieth century—stimulating them to construct a proper wind tunnel to produce more reliable numbers.[10] As a graduate student, I was unaware of any of this history—my whirling arm produced the same kind of systematic errors as had the classic ones. Fortunately, I discovered these errors myself and lost only a little time. What happens with whirling arms is that the air through which they pass is quickly set into motion, something all too easy to miss. I had reinvented a bad wheel. The aphorism of reinventing the wheel refers both to the history of technology generally and to the present subject all too specifically! As has been said, those who do not know history have to live it all over again.[11]

Lest anyone grab what has been said so far and run with the claim of a fundamentally irrotational natural world, of an unnaturalness to the truly rotational motion of our technology, the chapter needs several anything-but-parenthetical penultimate observations. Putting something into true rotation does happen to be trickier than you might think. Imagine two people sitting next to each other on well-greased swivel chairs. One gives the other a spin; the first spins in the opposite direction, canceling out any net rotation—canceling out at least in the sense of a particular physical variable called "angular momentum" (about which more later). Dip a paddle in water and push it broadside—you create two spinning vortices, but they go in opposite directions. It looks as if you have to step outside the system to initiate net rotation. The easiest way to do this is to stand on the ground and give the person on the swivel chair a push. Strictly speaking, you haven't really stepped outside but have just enlisted the aid of the earth, which has so much mass that the minuscule change in its spin can be ignored. The spins of severe storms, cyclones, are balanced by opposite spins, anticyclones, much larger and turning far less rapidly—and noticed for what they are mainly by meteorologists.

So rotation does occur and commonly, although typically balanced by simultaneous counterrotation somewhere else. The counterrotation just may not be obvious. Sometimes it's offloaded onto another medium. Thus a maple seed breaks from the tree and begins rotating as it descends. What counterrotates? In this case, an equivalent (again in terms of angular momentum) amount of air.

Nor are maple seeds unusual in their rotation. Equivalent behavior has evolved in seed-bearing plants on many occasions. One version (again with multiple independent instances) even does a double rotation. The descending seeds (properly fruits) of such trees such as ash and tulip poplar have an obvious—and aerodynamically odd—symmetry. They lack the leading and trailing edges of normal airfoils, including maple seeds. When dropping, they not only operate as auto-gyrating propellers but simultaneously generate their descent-slowing lift in the manner of a playing card that's flipped so it rotates lengthwise as it falls and as a result glides laterally as it drops downward. (The phenomenon of spin-induced lift underlies curving baseballs, sliced golf balls, and the like; the action long escaped notice in auto-gyrating seeds.) Thus—autorotation as well as auto-gyration, to attempt a linguistic distinction.

Nonetheless, unbalanced rotation—or rotation that for our purposes can be considered unbalanced, does happen—and does so under fairly ordinary circumstances. One of the less obvious but most basic phenomena in fluid mechanics is the way that the speed of a flow gradually diminishes as one gets closer to a surface along which it's flowing. At the surface itself, the speed of flow hits zero. So, on every surface across which air or water flows, there's a speed gradient. Put a lightweight sphere in that gradient on a surface, and it will be blown harder on top than on the bottom, and it will roll—

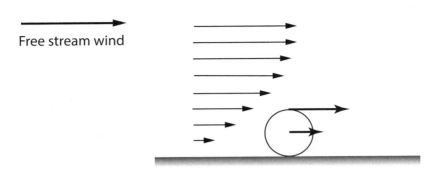

Free stream wind

Figure 1.10. Wind acting on a sphere near a surface will tend to roll the sphere along the surface—even without any friction between sphere and surface—because the speed gradient in the wind produces unequal torques on the sphere.

rotate—even if it doesn't touch the surface itself, as in figure 1.10. That's a good part of why tumbleweeds are called tumbleweeds— they tumble as they blow downwind even without frictional contact. with the ground.

And living nature provides a splendid contrast between circular motion with and without rotation. When I was a student, aspiring biologists were taught that nature had never invented the wheel— meaning, of course, the wheel-and-axle mechanism. If indirect evidence suggested otherwise, it was swept under the rug. As George Bernard Shaw said, "A good cry is half the battle," and science ceaselessly searches for robust generalizations. But one unequivocally rotary engine was discovered by Howard Berg and Robert Anderson in 1973—that of the bacterial flagellum, initially on that common creature *Escherichia coli* that we harbor by the millions in our intestines.[12] It's a rigid helix, driven, propeller-like, by a rotary engine located in the wall of the bacterial cell. By contrast, the flagella of nucleated cells such as our own, a lot larger and of quite different internal structure and chemistry, go around irrotationally—at least those that turn with a helical motion rather than limiting their waviness to a planar oscillation.

The overall playing field has now been defined—what about its length, the start and finish of the game? Here we're talking about human prehistory and history, a distinction that has always mystified this non-historian and non-anthropologist. While what follows traces no single sequential narrative, it covers a period that might start about 20,000 years ago, long enough so the earth sustained major climate and sea-level changes but not long enough for significant continental drift. Both our knowledge of the past and our knowledge base itself have accelerated with time, so the author needs ever more words to cover ever less elapsed time. That is, until a sharp deceleration roughly a hundred years ago: the era of the final triumph of combustion engines over muscular engines. Yes, we may be seeing some recession in those all-too-literally infernal machines, but one hastens to remind oneself of the large fraction of our electricity yet based on burning hydrocarbons. Just as most of our eggs still come from the supermarket rather than from the henhouse of a purveyor at a farmers' market, the thing on the roof remains far more likely a satellite dish than a solar panel.

[2]

Wheels and Wagons

Circular motion must have been familiar to the ancients. Never mind Eratosthenes' measurement of the circumference of a round earth, Copernicus's heliocentric solar system, and Kepler's elliptical planetary orbits. None were by any stretch your typical human of their times and, relative to the present account, even the first isn't all that ancient.[1] More immediately, stones sometimes roll, as do sections of tree trunks; better yet, both progress along the ground as they do so. The scheme, briefly noted in the last chapter, of using trunks as rollers beneath heavy objects certainly saw wide use. Marking the extreme were the peculiar predilections of several distinct prehistoric cultures for moving chunks of stone weighing many tons over considerable distances—for reasons not obviously utilitarian. On a more everyday level, pots are stirred, bundles (and infants) are wrapped, and so forth.

But the notion of two or four rotating wheels combined with half that number of axles remains a subtle concept; worse, the arrangement has no obvious counterpart elsewhere in either the living or non-living world. Sure, logs will roll beneath a large tree trunk or megalith, but achieving analogous rotational motion while fixing

the relative location of the rollers with axial rotating attachments—
axles—represents a considerable conceptual leap. It depends on recognizing at some level that, when a circular object rolls, it has a point
or line, the axis, that remains in a fixed place relative to that object.
Or, put another way, it depends on recognizing that the object can be
rotated around a point or line without going up and down. And then
it requires recognizing the advantage gained by connecting that line
or point to the rest of the vehicle in a way that permits reasonably
free rotation, in short, through a structural discontinuity. We call
the latter a "bearing," and it will provoke many more words further
along. A biologist, long immersed in evolutionary thinking, immediately wonders about—and then almost as immediately dismisses—
the possibility of intermediate, transitional forms between rolling
logs and wheeled vehicles. Or, put in slightly different terms, between assisted sledges and proper wagons.

Again, the crucial advance consisted not merely of a rotating
wheel, but of the combination of wheel and axle, one rotating with
respect to the other—or, put more simply, one rotating and the other
not doing so. In a real sense, then, it's the axle that defines the technological revolution.

And that's the component that, as far as we know, does not occur
as a propulsive arrangement in creatures larger than bacteria. The
biologist (aka the author) hastens to point out that rotation itself,
meaning repeated cycles and not just arcs, may not be common in
nature, but happens in a decent number of systems clearly not tied
together by close common ancestry.[2] Tumbleweeds appeared in the
previous chapter. One can add dung beetles, which make and then
roll spheres of mammalian dung larger than themselves, dung that
will be food and shelter for their grubs. Some animals can form
themselves into balls and roll on occasion—pill bugs (or roly-polies,
really isopod crustaceans rather than any kind of insect, let alone
true bugs) and armadillos, among others. Some do it habitually, notably a stomatopod crustacean that progresses when washed up on
a beach (where its legs are ineffective) by somersaulting back to the

water.[3] A coconut octopus can take advantage of the hemisphericity of a partial coconut shell, climb in, and roll around together with it.[4] The wheel spider of the Namib desert can roll down hills by doing cartwheels many times per second.[5]

Beyond the conceptual leap, producing any kind of useful two-wheeled cart or four-wheeled wagon presents awkward practical difficulties. Rolling logs experience little resistance to their movement. Resistance comes mainly from the lack of perfect rigidity of both themselves and the substratum and from the bumpiness of both, which mainly means irregularity of the substratum. Beyond energy losses from flexing their non-rigid and imperfectly elastic materials, functional wheels must endure sliding friction. That occurs either between a wheel and a fixed axle or between a fixed wheel-axle combination and the rest of the vehicle, as in figure 2.1. In a world of crudely shaped wooden artifacts, such friction could easily offset any advantage that might be gained by using wheels rather than carrying or dragging a load. Until quite recently we've taken advantage of the self-lubricating property of one variety of wood, lignum vitae, to make bearings for the driveshafts of propeller-driven ships, but

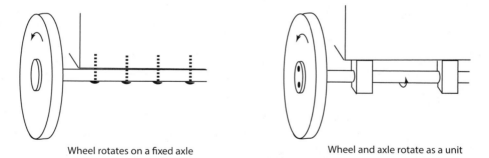

Wheel rotates on a fixed axle   Wheel and axle rotate as a unit

Figure 2.1. Two versions of the wheel-and-axle mechanism. On contemporary front-wheel drive cars, the rear wheels use the version on the left, where the axle is stationary and the wheels turn on a bearing; the front wheels use the version on the right, where the axle is fixed to the wheels and both turn as a unit.

antiquity knew nothing about that trick,[6] and lignum vitae is exceptional stuff. Metal inserts might have been used as bearings, and they have been standard fixtures of all recent wood-based, wagon-building cultures. But the combination of the limited availability of metals, the conceptual jump needed to use discrete bearings within otherwise wooden wheels, and the problems of fabricating axle-bearing assemblies would together have amounted to a stumbling block for use of what to us looks like an obviously superior bit of mechanics.

One way or another, the bearings of wheels demand lubrication; fortunately animal fat provided good-enough grease for the slowly turning ones of antiquity. But it needed frequent reapplication as a result of leakage and its own deterioration. Nowadays we make specialized bearings of metal—of hardened steel ball bearings or roller bearings, of porous bronze oil-retaining bearings, and so forth, or of particular plastics that have very low values of sliding friction such as Teflon. So lubrication remains a consideration, exacerbated by the heat released by rapidly moving, metal-against-moving-metal contacts. Old trains carried an oiler, a person supposed to keep track of when and where lubrication needed application; even so, seized-up wheels sometimes caused derailments. Not so long ago, one of the periodic service items for an automobile consisted of repacking the grease of front-wheel bearings. Highways showed occasional single skid marks arcing off left or right, places where some vehicle suffered a seized bearing.

Nor are bearings the only problem. Making wheels themselves is no trivial business. If one takes the rolling log as a mechanical starting point and by some means slices off a crosswise disk, that disk makes a particularly fragile and therefore minimally useful wheel. (Incidentally I don't mean to claim any specific technological sequence here; wheeled vehicles more likely evolved from dragged sledges than from rolling logs.) The grain of the wood will run in the worst possible direction, that giving the least resistance to fracture from the stresses of ordinary use. A tight band of metal around the periphery would be of great help—beyond the way it would minimize peripheral wear—but tight bands of tension-resisting metal were anything but ordinary items in the pre-classical world. A disk cut from a lengthwise slab of wood will do better, but its fragility will depend on where in the cycle of rotation it bears the load—when the grain runs up and down it will perform almost as poorly as a transversely sliced disk. And that slab would need to be a wide one, a slice of no trivial tree. Perhaps the simplest fix amounts to the invention of what we'd now call plywood. Put two or three longitudinally cut wooden disks (or sets of planks) together face-to-face, attached as best one can with glue, lacings, pegs through holes (treenails), or some other device. The result will be a decently functional wheel, if a counterproductively heavy one. We shouldn't be surprised that both wheels and their construction soon become specialized trades wherever wheels come into use, with spokes connecting hub and rim to keep weight down, with specific woods mounted in specific orientations, with a respected profession eventually giving, among English speakers, the surname "Wheelwright."

Another factor compounds this particular problem. For bearing loads, wheels of small diameter and large width should do about as well as ones of large diameter and narrow width. Making wheels of lesser diameter has to be much easier than making ones of greater diameter. Put another way, a low value of this so-called aspect ratio, outer diameter divided by tread width, makes for easy construction. At the same time, high values of the aspect ratio—big wheels—give much improved functionality. The world's surfaces are never perfectly smooth, and the ease and practicality of going over bumps depends on the height of the bump relative to the wheel's diameter. The bigger wheel runs more easily over a given bump—a bump half the diameter of a wheel forms a complete barrier, and a quarter diameter represents the practical limit of what you can ask of a real wheel. Even that lesser bump demands far more tractive force than does riding along on any smooth, hard surface. So, for all but the best of roads, wheels need to be of large diameter—not for nothing were the bodies of old wagons, carts, and carriages commonly mounted

Figure 2.2. A Murray Farm Wagon offered for sale in Catalog #13 of Montgomery Ward & Co. of 1875; note the extremely large (by modern standards) wheels. Large wheels are generally better on bad roads, where the large diameter helps smooth out bumps and allows the wheel to surmount relatively larger irregularities. The odd-looking appliance at the top is a spring-loaded seat.

between huge wheels rather than perched above small ones like those beneath the rolling stock of railroads. And the rougher the anticipated routes, the bigger the wheels, with notably large ones on the Conestoga wagons and then the prairie schooners that moved North American pioneers westward, not to mention ordinary farm wagons (as in fig. 2.2). Theodore Roosevelt, in his book about his nearly disastrous 1914 expedition through the southwestern Amazon basin,[7] mentions two-wheeled carts with wheels seven feet in diameter, drawn by six oxen—heroic extremes necessary to traverse the marshy ground near the upper Paraguay River.

Bear in mind this functional aspect of wheel size when you shop for wheeled luggage. Some models (I shudder at the currently faddish spinner luggage) have wheels so small that even rough pavement or the shortest and densest carpeting will ask you to pull appreciably harder. On tiled floors or bricked walkways, they'll add to the unpleasantness by passing more vibration up through their handles.

The shift from horse-drawn omnibuses to horse-drawn streetcars in mid-nineteenth-century cities provides a particularly nice case in point. Omnibuses differed little from freight-carrying wagons, and, like wagons, they had to take in stride all the bumps of city streets. So they needed large, sturdy wheels, not to mention very heavy axles. Streetcars—horse cars or, more recently, trolleys—ran on tracks. That not only permitted smoother, faster rides and demanded fewer horses (no trivial matter, as this last allowed lower fares), but they permitted a major redesign of the vehicles. Streetcars had wheels small enough to go beneath the floor of the trolley, which (with the absence of bumps) allowed shorter and lighter axles. Thus, the floor could be lower, greatly easing entry and exit.[8]

Yet more trouble. Put four reasonably aligned wheels on a body, and it goes straight ahead when pulled or pushed. Turning? One can guess that lifting or dragging the front to shift heading, while inconvenient, may not have been completely out of the question for the earliest wagons. At the least, at times even a non-steerable wagon should be usable. In particular, long stretches of straight path might well have been easy enough to lay out where wheeled vehicles traveled, somewhere within the Middle East and India. Still, inability to turn must have represented a serious drawback for full practicality, and it has been suggested that the first wagons may have served in religious processions rather than in ordinary transportation and commerce. Even now some Indian temples have their large "juggernauts" for ceremonial processions, and our word, appropriated for something both destructive and unstoppable, comes from that bit of Sanskrit.

So how to make a four-wheeled vehicle turn? The simplest scheme, conceptually at least, attaches the two front wheels with their axle or axles to a subchassis beneath the floor of the vehicle—as in the classic little red wagons we knew as children. Attached by a turning pin to that floor (a short, vertical axle, in effect), it forms a horizontal wheel-and-axle arrangement (fig. 2.3a). The pole or poles for pulling the wagon then extend forward from this subchassis so the wagon will follow in the direction of the force from whatever is

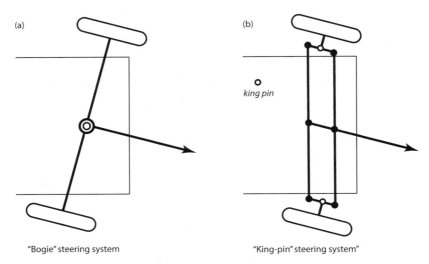

(a)

(b)

king pin

"Bogie" steering system   "King-pin" steering system"

Figure 2.3. Two steering systems. The open circles mark the locations where the steering mechanism is pin-jointed to the chassis. The closed circles indicate joints that permit turning of one member relative to the other. The scheme on the left benefits from simplicity but can easily become unstable; the scheme on the right is considerably more complicated but permits more freedom in positioning the wheels and maintains stability even in tight turns.

pulling. Turning that way raises two other problems. First, the sub-chassis has to turn with respect to the wagon's floor, and that joint brings friction back into the picture, now friction of one large plate against another. With that joint also comes potential wobble of one major component with respect to another. And second, the wheels now move back and forth and in and out as well as being redirected laterally when rotated for steering. So they have to be mounted farther outboard or made smaller to fit under the chassis, and the wagon becomes tippier when turning. Children pulling each other or coasting downhill in those little red wagons know this last problem quite well.

Better provision for turning consists of mounting front wheels, each with its own short axle, on a pantograph linkage of, minimally, four struts—two running crosswise and two lengthwise (fig. 2.3b).

With the pole or poles for pulling connected to the same linkage, wheels can turn with little fore and aft movement. But one vertical turning pin has now become four or six, and the connection between the wheel assembly and the floor of the wagon has become considerably more complex. Nonetheless, it has almost always been regarded as worth its complications (children's wagons and some flat-bed carts being exceptions). In particular, a pantograph linkage no longer demands that the front wheels fit under the vehicle's floor—at most a shallow wheel well suffices. Still, even though the arrangement handles the worst of the steering problems, it doesn't take care of everything that's relevant. When going in a circle or any arc of a circle, the wheel on the inside of the curve must turn more sharply—its path describes a smaller circle. At least for wheels, mounts, and bearings of decent strength running on roads of dirt or gravel, a little harder pull solves this particular problem. Wagons commonly minimize these problems by pairing large rear wheels with smaller front ones and moving both sets forward. Smaller front wheels reduce the room needed for lateral wheel movement when turning. And moving both sets forward make the rear wheels bear considerably more than half the load, so the misalignment of front wheels when turning makes less trouble.

Bear in mind that, when considering these wagons, we're talking about a serious technology. We tend to view the remains or reconstructions in museums as crude devices, perhaps because we no longer have proper eyes for wood-based mobile devices. In their later years, wagons were asked to carry remarkably heavy loads—up to eight tons for Conestogas—over the crudest of roads, drawn by as many as eight heavy draft horses.

Not that provision for steering must be incorporated for a vehicle to work at all. Wagons apparently enjoyed a long history before they could be readily steered. Roads either developed ruts or else ruts were deliberately created, and a wagon well stuck in its rut could get along with little or no deliberate driver-directed turning. This road-directed steering, you'll not be surprised to learn, had the secondary effect of fixing the distance between paired wheels. While values varied from place to place, most were between about four and five feet.[9] Given the stature of humans and of our practical draft animals, wheels around four feet—a meter and a bit—apart probably defined a reasonable track size for a wagon, with a somewhat larger width for the road it traveled. The overall size provides sufficient clearance for a pair of draft animals—perhaps a pair of oxen pulling on a central wagon tongue. The horse-drawn streetcars of the nineteenth century and the electric trolleys that supplanted them drew, if inadvertently, on a very ancient tradition. That ancient gauge has been remarkably persistent and pervasive. It has also determined the width of a single-lane road and the lane widths of multi-lane highways. Again they're a bit wider than the wheel gauge since vehicles often extend outward beyond their wheels and also need some space to allow for the vagaries of steering and operator expertise. The requirement of almost all wheeled vehicles for roads, together with their relatively fixed track width, is a boon to archaeologists. Evidence of roads constitutes good evidence of wagons—only hauled sledges and dragged megaliths might have left similar traces.

Even railroads, from the start riding on their own roads and thus lacking historical constraint, have stuck with that scale. Rail spacing has ranged from narrow-gauge mining roads, about two feet between rails but with protruding rolling stock, to the very wide seven-foot gauge adopted (and soon thereafter eclipsed) by Isambard Kingdom Brunel in the west of England. The United States and most of Europe use a gauge of 4 feet 8½ inches, one that traces to the English railroads of around 1830, but other gauges exist elsewhere. Russia and adjacent countries have tracks spaced just under five feet apart—initially a defensive measure, just enough to impede easy entry into the bloc.

A much simpler solution to the steering problem exists, although one unlikely to antedate four-wheel wagons. Just use a single pair of wheels, side by side, rather than two pair, preferably with bearings between wheels and axle rather than axle and body. Going from a draft animal dragging a load whose front the animal holds off the ground to such a cart requires nothing more than tying a wheel and axle to the back of the load. That may well have been the path taken in the evolution of wheeled transport, although the archaeological evidence suggests that four-wheeled wagons appeared first. But while two-wheeled carts were secondary inventions, and while carts can carry much lower loads than wagons, carts do have a long and colorful history with lots of diversity, from the war chariots of the ancient Greeks through the minimalist carts of present harness racing, with too many named versions between to be worth tabulation here, much less describing.

As an example of the best of utilitarian, non-military carts, I'll cite the so-called hansom cab, once common in cities, beginning with London, in the second half of the nineteenth century. The name alludes to its inventor and patent holder, Joseph Hansom, who made the first one in 1834. Most carts held two riders, but the hansom cab held three, two passengers side by side and a driver behind and above, who wielded reins and whip across the top of a closed cab, as in figure 2.4. It was light enough to need only one horse, and it was famously maneuverable in the chaotic urban traffic of its era. The large wheels minimized discomfort to all four (don't forget the horse!) from bumps and potholes and, presumably, would have let it run up and over curbs when convenient or necessary. At the same time, the center of gravity was low, reducing the chance of overturning; one version even allowed the driver to adjust the fore and aft location of the load's center. Arthur Conan Doyle's Sherlock Holmes and Dr. Watson were forever dashing around London in hansom cabs—his contemporary readers needed no explanation of any what and why for the vehicles.

But since it won't remain upright and level if left on its own, a cart can be loaded (and unloaded) only after hitching or with provision of a temporary front support. It lends itself poorly to multiple draft animals harnessed side by side, even if we see four-horse chariots depicted on some Greek amphoras, and it doesn't lend itself at

Figure 2.4. A Hansom cab and two cabbies, one reduced to a pedestrian state (from Thompson and Smith, 1877). Note the large wheels (to improve the ride) and low position of the cab (to minimize the chance of overturning). Only the driver is located high above the center of gravity, where he has an unobstructed view.

Figure 2.5. One of several drawings in Robert Thurston's (1894) careful discussion of how to best load a horse-drawn cart—sadly published only shortly before their displacement by motorized vehicles.

all to harnessing animals in series. The load has to be carefully positioned over the wheels, as we can see in an illustration (fig. 2.5) from Robert H. Thurston's 1894 monograph, *The Animal as a Machine and a Prime Motor*.[10] Offsetting these disabilities, carts enjoy better speed and maneuverability.

Thurston's book has much more to tell us. It represents the best American engineering in the last days in which animals remained crucial for land transport. At least in that particular guise, animal biomechanics then held far greater practical importance than any time since. Thurston organized the departments of mechanical engineering at, first, Stevens Institute of Technology, in New Jersey, and then Cornell University; he was also the first president of the American Society of Mechanical Engineers, and he somehow found time to write a remarkable number of still-readable books. The data he cites for the cost of animal-powered transport remain reasonable figures. They give us a good idea of the relative value of wheeled over non-wheeled load carriage.

But before turning to his numbers, I should put in a reminder of the difference between work and power. *Work* is how much of an energy-demanding task is done, while *power* is the rate at which it's done. You can do a big job if you work slowly and take plenty of time; the same job done more rapidly might leave you sweating or breathless—your power output has increased even if the total work has not. Work is measured in units of foot-pounds (how far times how forcefully) or calories or joules (if you're a physicist) or watt-seconds (or kilowatt-hours on your electric bill; one joule equals one watt-second). Power has units of foot-pounds per second or calories per second (often kilocalories per hour or day) or watts (joules per second).[11] Reading the pages that follow doesn't depend on remembering the specific content of the previous sentences—just the distinction between an amount and a rate. Bearing in mind that distinction, a few figures:

- A man carrying a load of optimal weight on level, hard terrain and returning unloaded can move that load for 150 foot-pounds per second (200 watts); he can do this for 6 hours a day. If provided with a two-wheeled barrow, he can move it for 370 foot-pounds per second (500 watts); this second task he can do for 10 hours a day. That's 2.4 times faster, with four times more accomplished over the course of a day.
- Similarly, a horse walking while carrying a load manages 970 foot-pounds per second (1,320 watts), but when using a cart to carry the load at the same speed, it can achieve 5,400 foot-pounds per second (7,300 watts), 5.5 times more.

Not that these figures should be assumed directly applicable to early wagons and carts, traveling on very poor roads or paths with wheels that were compromised by crude wood-against-wood bearings. Thurston necessarily considers late nineteenth-century contemporary practice about as good as it got. Nonetheless, the early vehicles must have been viewed as advantageous (or, less likely but not ruled out, valued by the local deity), but their advantage could not have been so overwhelming.

(Again, someone familiar with the relevant physical and biomechanical science may object that, when going horizontally, no net physical work is done. Just because the units fit doesn't mean we're dealing with power output in a strict physical sense. Underlying these data are estimates of pulling or pushing force for the carts, with an equivalent force assumed for directly carrying loads. Force times distance gives work; again, dividing by the time taken puts the matter into terms of power—however odd the implied definitions of work and power. In practice, these estimates provide good comparisons among themselves but not to most other situations.)

Another view of just how good wheeled transport can be comes from comparing a walking or running human with a vehicle where power efficiency represents the most important consideration for its design. (For walking and running, the cost per distance doesn't vary much with gait or speed, unlike the cost per time.) I am referring, of course, to bicycles, evolving only slowly now after almost two hundred years of refinement. Running at 9 miles per hour (a little better than 7-minute miles) takes 3.7 times as much power as pedaling at that same speed—despite cycling's requirement that 25 or so pounds of equipment be attached to the body.[12] The comparison bears obvious relevance to the shift in human-powered transporters from traditional rickshaws that depended on a single human walking or running to ones in which the driver turned pedals as if riding a bicycle. And it also explains the dependence on pedaling of all modern designs of human-powered aircraft.

So applaud the magnificence of the invention of wheeled wagons and carts, perhaps the most important single mechanical achievement of human technology. But at the same time recognize that it couldn't have happened in a single burst of creativity, that even the simplest wheeled vehicle is better regarded as a system than it is as a single device. And prerequisite are sufficient development of wood-working technology, of draft animals and techniques for harnessing them, and of both the ability and willingness to produce roads.

With all this talk about historical (or prehistorical) antecedents, something should be said about the specific antiquity of wheeled wagons. Both traces of roads and images of wagons help us assign dates. Archaeological evidence has roads existing by between 3500 and 4000 BCE—1,000 to 1,500 years before the Great Pyramid of Giza, in Egypt, to give a reference point. Where? Probably the Middle East, as mentioned, but with reasonable evidence of early wheeled vehicles from Britain to India.[13] Do the widespread sites indicate diffusion or independent innovation, a favorite point of contention among archaeologists and anthropologists? I'd opt for diffusion, even against my prejudice as a biologist who's impressed with the inde-pendent origin and convergent evolution of innumerable features of organisms. A thousand years to diffuse perhaps 4,000 miles asks only 4 miles per year, scarcely more than an hour's brisk walk. And we're talking about a device that by definition increases human mobility.

Quite a different line of evidence, this from linguistics, agrees with that date of 4000 BCE. The common roots of modern Euro-pean and South Asian languages can be traced back to what is called "Proto-Indo-European"; in particular, all these languages use one or another derivative version of the word for wheel in that hypothetical ancestral tongue. So inhabitants of the steppe region north of the Black and Caspian Seas (modern southern Russia) must have had the device, or else why would they have needed a word for it? Not that linguists assert that wheels were invented there—just that they were known there and then.[14]

Still, a downside of wheeled transport lurks—in fact, several downsides—to muddy the metaphor. We've already talked about the difficulties of making wheels, of providing them with bearings, and of steering the resulting wagons. The requirement for roads needs additional examination. With the exception of wheelbarrows and, quite recently (and only recreationally), mountain bicycles, wheeled vehicles travel on roads. These may not be graded and paved, but they're more than simple footpaths or arbitrary cross-country headings. Under most circumstances, laying out and maintaining roads takes deliberate effort. And so their presence or remains implies some minimal level of both population density and civil authority. It can't be entirely accidental that some of the earliest evidence of carts and wagons appears where agriculture had long been established and where irrigation was practiced—the combination both permitted and required higher population density. Wagon-dependent cultures have always devoted considerable effort to road making. Soft or swampy areas have often been crossed by "corduroy" roads, named after the fabric from which we make corduroy trousers and consisting of continuous surfaces made up of tree trunks lying side by side, crosswise to the path of the road. Unless embedded in especially acid soil, corduroy roads deteriorate rapidly, and any wagon traveling on one gave a ride that was both slow and bumpy. But even that was better than a cleared path through a marsh.

A good road must not deform much when it feels the presence of a wheel. That's why wheels rolling on rails, even rails of wood, have been preferred where the task required repeatedly carrying heavy loads along exactly the same path. That's also why bringing ore out of mines made early use of railed transport, with the wagons, individually or as trains, powered by animals, including humans. And that's why the earliest successful steam-powered vehicles for land transport rode on rails, not roads. Metal wheels on a metal (or metal-surfaced) rail ensured minimal deflection of both wheel and path, keeping friction low enough for heavy, underpowered trains. Even now trains always accelerate gently—they're not especially powerful relative to their weight. So, as implied when talking of corduroy roads, surface quality matters, and much of the history of road construction centers on the design of surfaces that are both hard and durable. Contemporary government rules disallow covering public areas with pleasantly soft carpeting—lest pushcarts and wheelchairs be nearly immobilized. Roman roads were famously good, with the imperial imperative of traveling relatively rapidly with the least number of draft animals. And we preserve the name of one of the greatest innovators in the road-surfacing business, John Loudon McAdam (1756–1836), although our present asphalt topping was introduced long after his time. Of course, "macadam" may be partly chauvinism and partly because the name Pierre-Marie-Jérôme Trésaguet does not so easily jump from printed page or screen to Anglophone tongue. In any case, "tarmac" does less violence to history.

A few data will give some idea of the variation in the force required to pull wheeled vehicles along a level path on different surfaces, here given relative to the weight of the loaded vehicle. The last column gives values relative to the last datum, the gold standard, one might say.[15]

- Loose sand, iron (rigid) tires   0.143   35.8
- Broken stone on earth, iron tires   0.0286   7.2
- Good surface, pneumatic tires   0.0167   4.2
- Brick road, iron tires   0.0111   2.8
- Steel-plate trackway, iron tires   0.004   1.0

No accident, then, that litters (sedan chairs, palanquins, etc.) with two or four bearers preceded rickshaws powered by a single person—even for use on level ground.

By contrast, a marching Roman—or any other—army demanded far less of the surface over which it traveled. In terms of required power, legged locomotion may be far less efficient than wheeled locomotion, but it's far more versatile. That's the main impetus behind efforts to develop legged vehicles and robots, despite their obviously greater mechanical complexity. The US military has expended large amounts of money over many years in attempts to produce legged vehicles, even ones of the limited efficiency and high cost tolerable for their uniquely non-commercial mission. As yet no design has been judged suitable for large-scale production and deployment.

Wheeled vehicles hitched behind draft animals suffer one further limitation. The bigger the animal, other things equal, the more it costs to climb a hill relative to covering level ground. Tests on treadmills have shown that a grade that costs a mouse 1.2 times its level best will cost a horse 7.3 times more.[16] And that's before we ask that the horse pull anything. The more nearly level the road, the greater the load a horse can manage. On a good road that crosses level terrain, a horse can pull a vehicle of about three times its own weight for extended periods, meaning roughly 3,000 to 4,000 pounds for a draft animal. On asphalt pavement, a horse asked to ascend a 3 percent grade can pull only 18 percent as much—540 to 720 pounds. On a grade of 6 percent, the horse can do only 6 percent as much as on the level—a trivial traction of 60 to 90 pounds. Bear in mind that the weight mentioned is that of the vehicle, not the force exerted by the horse. A good rule of thumb for force exerted by most draft animals is about a tenth of their own weights. When Theodore Roosevelt, talking of immigrants, suggested that a good citizen "should be able and willing to pull his own weight," he was, strictly speaking, asking the impossible.

Slopes of up to 6 percent, incidentally, are permitted for the US Interstate Highway System; think of the tax such a slope imposes when you're ascending behind a heavily laden truck. Of course, the better the road, the greater the relative advantage realized by making it level, part of the impetus for the slope-limitation rules set when the US Interstate Highway System was established.[17] One recalls the heroic efforts we've made to ensure minimal slopes for railroads—all those tunnels, embankments, and bridges. They pay off now in terms of energetic efficiency; originally they responded at least as much to the limited power-to-weight ratios of early locomotives.

That brings up a curious distinction between Old World and New World cultures. Wheeled vehicles spread throughout the first but remained unknown, or at least unused, in the second. The nearly complete absence of utilitarian metallic materials and of wheeled vehicles represents the two greatest technological differences between the cultures of the two hemispheres. Both must have had causes (barring the unlikeliness of pure accident), and both unquestionably had consequences. Other comparably challenging technological feats such as spinning and weaving seem to have been developed quite independently.

Often considered causative is the lack of appropriate draft animals in the New World. It lost its horses at the end of the last ice age and did not reacquire equids until escapees from the conquistadors went feral in western North America. The New World relatives of the camel, llamas and guanacos, are too small to pull vehicles of appreciable size, even if llamas do very well as pack animals. The most widespread bovid, the bison, has never been sufficiently tamed to be trained for service as either a draft or pack animal. Caribou (as reindeer) have been pressed into draft service in Scandinavia, but in North America they occur only in boggy areas that support sparse populations of humans. Some minor use of dogs has been reported, either dragging light pole-mounted loads or pulling sleds on the peculiarly low-friction surfaces of ice and snow. In neither instance would introduction of wheels had an obvious payoff. Canids are too small and, as carnivores, too expensive to provide with fuel.

Another possibility is that in the New World the most highly developed cultures, ones with substantial population densities, developed in areas not particularly amenable to wheeled transport, either, as with Maya of Central America, because of the density and surface sponginess of rain forest or, as with the Inca of Peru, because their domain was too hilly. The Inca did develop long trails, but these are perhaps better described as good paths rather than as roads. The great Inca Royal Road ran 3,500 miles, from what is now Quito, Ecuador, to Santiago, Chile. It had embankments and suspension bridges—and slopes that ruled out any vehicular use. Again, even the gentlest of slopes greatly taxes draft animals, and even if forward motion of a vehicle remains possible, it can be done only with a greatly reduced payload.

Finally, I take seriously Jared Diamond's suggestion that geography on the largest scale provides the crucial difference. He points out that Eurasia runs east and west, with accessible land connections from the Iberian to Korean peninsulas.[18] Similar topography—open and fairly level country, semi-arid land with resources for irrigation, similar suitable crops—gave a good chance that innovations might spread. By contrast, the Americas run north and south, with drastic changes in climate and topography along that axis. So something that might be advantageous in one place might be of limited practicality in most other not-too-distant locales. In other words, in the Old World, aspects of culture may have diffused more easily than people could migrate; in the New World, movement of people faced lesser impediments than diffusion of many important components of culture.

It seems unlikely that the relatively more recent arrival of dense settlements and agriculture in the Americas could underlie the absence of wheeled vehicles. Human populations were denser than apparent to the newly arrived Europeans, who met societies in the immediate aftermath of devastating epidemics of novel diseases. Agriculture? A land dominated by hunter-gatherers wouldn't have domesticated potatoes, corn, tomatoes, red peppers, squashes, many beans, quinoa, and other now-familiar fruits and vegetables. Finally, at least the Maya clearly knew about the possibility of rolling around on a set of wheels and axles. The evidence comes from, of all things, what appear to be toys of a kind still familiar. Clay animals had a wheel on each of their four legs, as in figure 2.6.[19] No toy wagons, of course—their presence would tempt us to presume cultural diffusion around the Pacific Rim from the Orient just as their absence argues against such diffusion.

Four-wheel wagons, two-wheel carts, occasional tricycles, and further vehicles with wheels in greater multiples of two—do these exhaust the range of historically interesting vehicular applications of the wheel and axle? One other must definitely be considered.

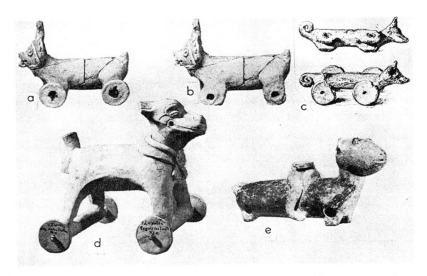

Figure 2.6. The famous wheeled toys enjoyed by presumably unimpressed Maya children (from Ekholm, 1946, © Society for American Palaeontology; used with permission).

You can create a vehicle with a single wheel, producing not just an impractical unicycle, but a fine wheelbarrow. It not only needs no particular provision for steering, but it asks far less preparation and maintenance of its route—a real road is unnecessary; a narrow path suffices. Wheelbarrows first appeared in China, about a millennium earlier than their adoption in the West, and they have been most elaborately developed there.[20] One can imagine an initial application that put a wheel on the end of some long load that was being dragged along by a single person or an animal as small as a dog. But the advantage over simply skidding along would have been slight, and we have no evidence of such an origin. Even now a wheel on a trailing load is rare. I do know someone who put a small swiveling caster on the top spire of an umbrella so he could pull it along behind when the umbrella was not in use—and attract curiosity when it was. In any case, the wheel of all respectable wheelbarrows normally graces the front.

In China, wheelbarrows enjoyed a long history as people carriers. In one multi-passenger design, equal numbers of people sat on each side of a wheel of huge diameter, thus keeping the load well centered over the wheel (fig. 2.7). Atop the wheel, anticipating modern carriers, was a small luggage rack. Somehow one person managed to push the thing, which says something about the smoothness, hardness, and level of the roads—most likely city streets—on which it traveled.

Wheelbarrows appeared in Europe about a thousand years ago, apparently derived from those Chinese ones rather than independently evolving. One wonders about the late arrival of something so ordinary and utilitarian, a cart that needs no road, one that can traverse a narrow plank and that can go cross-slope without tipping if guided by a human—and wheelbarrows have almost always been

Figure 2.7. A Chinese wheelbarrow; this particular one might be considered a muscle-powered minivan, one with a rear engine and front drive (from King, 1911, via http://www.lowtechmagazine.com/2011/12/the-chinese -wheelbarrow.html).

small things pushed by a single person. For substantial distances, at least where a road is available, a cart is preferable because of its near-perfect immunity to any sideways tipping that might result from in-attention or load shifting.

Wheelbarrows, though, face an awkward problem of design. In the real world, for reasons that by now should be obvious, the bigger wheel moves more easily than the smaller one. Bigger means heavier, and any weight penalty becomes serious when the engine consists of a single human and slopes might be encountered. A person can push or pull, as a generous estimate, about 30 pounds, a datum that has to be reduced considerably if the task must be sustained for very long.[21] At the same time, a load can be moved most easily if its center of gravity lies over the point of contact of wheel and ground rather than ahead or behind it—that central location means that its opera-tor need only push or pull rather than waste output lifting. American wheelbarrows have their wheels forward of the trough and require considerable lifting; old European ones were worse.[22] I noticed that wheelbarrows in Taiwan had their wheels tucked back closer to the bottom of the trough, although still in front of the center of gravity for any normal load.

A few years ago, I built a Chinese-style wheelbarrow based on an old bicycle wheel. A simple platform above the wheel made the thing too tippy for any use I could imagine. So I rebuilt it with two com-partments that flanked the wheel. Pushing and pulling on level pave-ment was startlingly easy—one mainly applied force to accelerate or decelerate. But the deep, narrow compartments limited its utility, and I found it useful only for moving around the leaves I raked in the fall. They packed in quite well but never taxed its load-carrying capacity enough to give it a real test.

Two more pieces of this business of going from place to place while perched above one or two wheel-and-axle mechanisms.

First, the data earlier referred to horses, mainly because those figures seem to be the most accessible and reliable ones. By the nineteenth century, horses dominated the draft animal world, at least in developed countries. So horse performance received the lion's share of serious attention—as one recognizes in the word "horsepower" rather than, say, "oxpower." But among the draft animals of antiquity, horses played a secondary role. They mainly served royalty and the military—both notoriously cost-insensitive.

And cost presented their main limitation. Horses had a long history of domestication that made them tractable enough for draft use. They may have required a more expensive diet than the main alternative, oxen; offsetting that, they went about twice as rapidly while pulling about the same load (again, Thurston [1894] provides fine data). But that equivalence could not be realized until the arrival of a proper equine harness. Oxen and other bovids-of-burden such as water buffalo and yaks have a fatty hump atop their thoraces. A cylindrical beam crossing in front of the hump can connect to a pair of rear-directed poles such as a pair of tongues extending back to the front end of a cart, as in figure 2.8a. Or a pair of oxen can share such a yoke, with a single wagon tongue running back from the yoke between them.

Horses and other equids such as donkeys, mules, and onagers lack that hump. Instead of the yoke bearing on the thoracic vertebrae and thus directly on the axial skeleton, the loop beneath the yoke will take the strain. The only practical route for that loop takes it around the neck of the animal (fig. 2.8b), so pulling hard chokes the unfortunate creature. Ancient Mediterranean cultures, nomads on the Eurasian steppes, and others kept (and ate and even worshipped) horses. But good draft animals they were not, so plowing and hauling depended almost entirely on bovids, with the occasional assistance of camelids.

What changed the picture was the arrival in the West and adoption of a proper horse harness, one with a padded collar—as in figures 2.4, 2.5, and 2.8. It seems to have come to early medieval Europe from China, where a series of improvements, beginning with the replacement of a yoke and throat strap by a chest strap

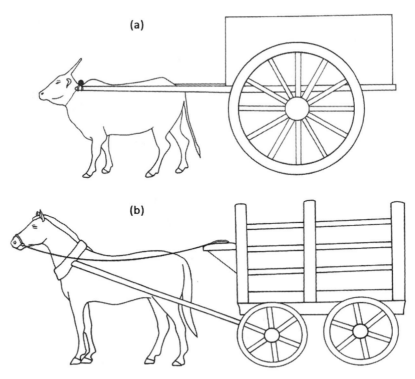

Figure 2.8. Two fundamentally different harnessing systems—the yoke, suitable for bovids, and the collar, suitable for equids (from Vogel, 2001). Bovids have a convenient hump to transmit forces to the load, but horses do not; a horse collar is a sophisticated device that allows a horse to pull a load without choking itself.

(fourth-century BCE) and ending with a recognizable collar (third-century BCE).[23] Thereafter equids as well as bovids could exert forces that reflected their full muscular ability as well as power outputs that took advantage of their full (and remarkably great) cardiovascular capabilities. In addition, because a rope (or trace) rather than a pole could connect collar and wagon, horses could be harnessed in long teams. In some places and for some tasks, equids were better; under other circumstances, one or another bovid was preferable. Horses gave better speed; oxen pulled harder at near-zero speeds. Horses like some oats or other enrichment to their feed; oxen, as ruminants, found grass a fully adequate fuel.[24]

The case for the crucial role of the horse collar in the history of Western technology specifically and Western culture generally was made forcefully and effectively by Lynn White (1907–1987), one of the greats of the history of technology. He asserts (and I have no grounds for disagreeing) that the arrival of a series of economically transformative inventions in the medieval period provided the social and economic latitude for the Renaissance. Besides horse collars, he points to horseshoes and stirrups, plus many other non-horsey items that, for instance, permitted efficient agriculture in the heavy soils of northern Europe.[25] For the next millennium, horses continued their role as our predominant non-human power sources. Nineteenth-century cities were possible only with fabulous populations of horses, and early cars and trucks were seen as a liberation from all the problems presented not by human but by equine population density—all those horses that, among other tasks, pulled wheeled vehicles.[26]

I ought to note, in passing, that the wheels of a vehicle can serve functions beyond mere conveyance; with a draft animal as motor, the only places for a power takeoff are wheels and axles. So those revolving wheels constitute a potential source of power—not free, since the draft animal requires fuel and maintenance—but conveniently accessible. The main limitations are, first, the obvious one of the maximum power output of the animal as harnessed, and, second, the friction of wheels with substratum, the resistance the combination offers to slippage. Boys of my generation, meaning youths of and before the mid-1950s, had an uncomfortable familiarity with the way power could be bled off turning wheels. We pushed reel lawn mowers; our pushing turned their wheels; and the turning wheels then spun the blades, which swept past and just cleared the cutter bar. You knew all too well if the grass was too thick, if the blades contacted the bar, or if lubrication was insufficient. Even under the best of circumstances, these machines provided a good—one maybe shouldn't say "sorely missed"—workout.

Figure 2.9. The "Q Drum" toroidal water tank—a wheel with a rope axle (see www.qdrum.co.za).

To bring us full circle (can one write without verbal allusions to rotational motion?), I can't resist describing a thoroughly practical human-powered wheel-and-axle vehicle that combines the utmost in primitive simplicity with modern materials, plastics in particular. In rural areas of many third world countries, water must be carried in containers from wells, lakes, or streams located at some distance from dwellings, a task that consumes considerable energy, usually exerted by local women and children carrying containers. At the headquarters of a humanitarian foundation in Little Rock, Arkansas (Heifer International), I saw a plastic toroidal water tank that rolls along the ground. A reviewer of this book tells me that it's a commercially available South African product called a Q Drum, the "Q" bearing a decent resemblance to its cross section.[27] As in figure 2.9, one person pulls the tank as a whole-vehicle wheel with a loop of rope that passes through the axial hole in the torus. That rope, perhaps greased with a bit of fat, acts as axle. But, with my long-standing interest in fluid dynamics, I'm left wondering how such a liquid-filled tank feels and how much work it takes to keep it in motion on irregular ground—especially when it's neither fully filled nor empty. Perhaps I'll have to build a model—as is my wont—or buy one.

The Q Drum bears an odd resemblance to a scheme that saw use in classical antiquity, but was used there and then for moving solid cylinders rather than liquid-filled tori. Short, square-sectioned elements of heavy quarried pillars were encased in wooden cylinders to form drums that could roll. Wrapping a rope around the outside allowed a team of oxen to power that rolling, at least until the rope ran out and had to be wound again onto the drum. Vitruvius, the author of one of the few surviving Roman technical treatises, attributes the device to one Paconius, who learned the hard way that they did not take kindly to steering.[28]

One should note in passing that this scheme makes the cylinder provide some leverage. By pulling at its periphery, the puller must pull twice as far as if pulling on the axis (here a virtual axle) but need pull only half as hard. I found the trick handy when rolling a 3-foot-by-20-inch oak log up a ramp into a pickup truck. Fresh oak weighs a little under 40 pounds per cubic foot (600 kilograms per cubic meter), so between a 4:1 slope of the ramp boards and that 2:1 rope advantage, my actual pulling force was only about 30 pounds.

Little note has been taken of a major problem introduced by any wheel-and-axle mechanism. So important is the matter of friction, the work-absorbing and heat-generating force where one surface moves across another, that it will form a major part of the next chapter.

Turning Points—and Pots

By some accounts, the potter's wheel antedates the wheeled cart, but the question of the relative antiquity of the two wheel-and-axle devices remains an open one. Also uncertain is whether either can in any real sense be regarded as the technological antecedent of the other—the specific inspiration, so to speak. Both may constitute devices based on the wheel-and-axle combination, on something that rotates in an otherwise non-rotating context. But they look different, perform different tasks, and, right from the start, present quite different technical problems. In particular (not to pretend suspense), a potter's wheel only minimally stresses its equivalent of the bearings of wheeled vehicles, counting on those bearings for little more than alignment of its vertical shaft—not that the requirement is trivial or easily done, though. On the other hand, the entire weight of the rotating parts of a potter's wheel presses on the axial end of the rotor, asking that the device provide a so-called thrust bearing of real competence.

The last chapter mentioned the problem of bearings but made little of it except to note the ancient trick of applying animal fat to grease the skids, the wood-on-wood sliding surfaces. That's the basic task of a bearing, the subject of what may initially seem something of a digression. When one surface slides across another, keeping it going takes a force to offset what in technical terms is called the "shear stress" of the interacting, moving surfaces. (Shear stress will reappear and play a major role when we talk about spinning fibers in chapter 9.) The bearing has to ensure that adjoining surfaces that move with respect to each other do so with as little friction—with as low a shear stress—as possible without losing contact (or near contact) with one another. Since a potter's wheel puts a different spin (sorry!) on the bearing business, perhaps we should take a closer look at the basic issues and solutions—get our bearings, one might say, in a sense more literal than mere compass directions. Contemporary technology runs on one or another of many thousands of commercially available bearings—a look at a catalog such as the online one of McMaster-Carr provides a look at the common varieties and an impressive illustration of their diversity.[1] To appreciate wheeled technology, it's worth a minor digression into friction and bearings.

A bit of bother does come from the less than fully logical divisions and nomenclature for these quiet heroes of modern machinery. In particular, one can easily get confused between names referring to the construction of bearings and names referring to their uses. (For the biologist to complain about some area's excessive focus on classification smells of a pot calling the kettle black.) For our purposes, we can recognize the distinction between bearings for linear motion and those for rotational motion. The first won't be of present concern, even if they're everyday items—we're all too aware that heavy drawers, as on file cabinets, either have or else badly need specific bearings; they rarely run well on wooden sliders.[2]

Rotational bearings deal with either or both of two loading regimes. Bearings on the wheels of carts and wagons—rotary shaft bearings—encounter almost purely radial loading, that is, loading at right angles to the long axis of the axle (fig. 3.1a). Bearings on potter's

(a)   (b)

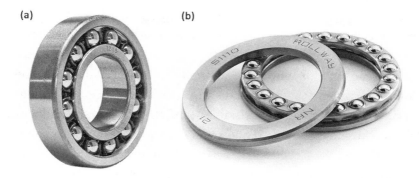

Figure 3.1. (a) A set of ball bearings designed to resist lateral (radial) force;
(b) a set of ball bearings designed to resist axial forces. Note the orientation
of the steel balls relative to the race in each case—the trick to reducing friction is to permit the balls to rotate freely at right angles to the load.

wheels face almost as pure a loading regime, but this time in an axial
rather than radial direction (fig. 3.1b). A good potter's wheel has to
be heavy, for reasons that will become clear shortly. Since it presses
down on its fixed support, the resulting friction can be reduced only
by a thrust bearing.

So how can one build a good bearing or, better, a pair of bearing
surfaces? One way that sometimes works (depending on the kind
of lubrication provided) consists of minimizing the contact areas in
such a way that the speed of one contacting surface remains as close
as possible to that of the other. For sleeve bearings, where a shaft
turns within a close-fitting housing, that means using the skinniest
of shafts. Each rotation of a wheel requires sliding each point on the
circumference of the shaft once around within a fixed housing, and
the circumference remains a fixed multiple (the constant, of course,
is π, or pi) of the diameter. Obviously, a lower diameter means a
lower relative speed of points on the shaft against the bearing surface
on the housing. That, in turn, means less sideways loading (shear) on
the lubricant, if one is used; it thus means less buildup of heat since

the power lost in friction depends on the frictional force times the speed of movement, and power loss rate translates into the rate at which heat is generated.

But the fly in the ointment: naturally, the skinnier the shaft, the lower its stiffness (both flexural stiffness, or resistance to bending; and torsional stiffness, or resistance to twisting)—and the relationship is extreme, as some numbers will illustrate. Halving the diameter of the shaft will reduce contact speed twofold, but it reduces the shaft's stiffness no less than sixteen-fold. According to a standard handbook, under conditions where a 3-inch (diameter) shaft can transmit 54 horsepower, a 1.5-inch shaft can transmit only 7 horsepower.[3] Clearly the option of making a skinnier shaft bumps into severe limits!

On a small scale, meaning for the most minute of loads, a particular solution has enjoyed several centuries of both good press and practicality. While out of the question for wagon wheels, sometimes one can taper the end of an axle down to a sharp point and arrange matters so that the point turns in a pivot hole in something harder than a metal. That minimizes contact area without excessively increasing the rate of wear—wear that would lead to both wobble and friction. The bearing surfaces of choice have been minerals, most often rubies or sapphires—originally naturally occurring gems, now mainly synthetic forms of aluminum oxide. Friction is low, and lubrication, while often used, isn't always necessary. Mechanical watches made much of the number of jewels in their movements—almost always an odd number of jewels since the oscillating component, the balance wheel, used only one.

In fact, jewel bearings have a wide range of applications beyond fancy watches. Good mechanical compasses use them, as do the better electronic meter movements of the ever-less-common swinging-needle sort. Some tiny motors and gyroscopes depend on them as well. Jewel bearings were invented in 1704 by a Genevan and two Frenchmen living in England; in a few decades, they became regular features of the best English clocks and watches. Oddly, jewel bearings were not used by continental timepiece makers until the end of that century. They gave good service in John Harrison's deservedly famous marine chronometers of the early eighteenth century. Harrison's instruments were good enough for seamen on extended voyages to tell their longitude with a high degree of confidence from astronomical observations. For latitude, one may need no more than, for instance, the elevation of the North Star, Polaris; but for longitude, one needs the time of your measurement, hence the critical role of the chronometer.[4]

An equivalent difficulty with contact area faces thrust bearings. One can reduce contact speed by sharpening the point of contact. Halve the diameter or radius of the contact, and you halve the speed at its periphery. But at the same time you've also quartered the area of contact (remember, $A = \pi r^2$). That, in turn, quadruples the stress, the force per area, that wears down the bearing surfaces, dulling any sharp point. Still, area reduction shouldn't be dismissed out of hand—it has its place and has the virtue of simplicity. Just be sure to start with a nicely stiff material, perhaps hardened steel.

A second consideration, typically at least as important as relative speed and area, is the sliding friction between the two surfaces that the load is pressing together. One simply can't talk about bearings without saying something about friction, but at least much that matters about friction follows one's normal intuition. Friction, to put it in context, is a dissipative process, one that by its very nature generates heat. Forcible deformation of a solid—as when you bend or twist a piece—is another. Yet another is viscosity, the analogous process in liquids and gases. Rub two surfaces together and heat appears; stir a liquid vigorously enough, and it will become noticeably warmer—syrup spun in a blender will provide an example for the skeptic. Friction, unusually, occurs at the interface between pieces; deformation and viscosity happen within solids and fluids, respectively. That makes the ability of the material to conduct the heat away of particular importance for sliding friction. Metals are quite good, while wood and plastics are poor. Heat typically changes the properties of a material; the heat released by sliding friction can change the relative sizes of the sliding parts or cook out any lubricant. Friction is almost always bad—although we do enlist its aid in stopping our cars.

Not surprisingly, friction varies widely. Again, occasionally we want high friction, as with shoe sole against floor or tire against pavement, but at other times, as with most of the present systems, we want to keep friction as low as possible. At this point, we need some numbers to give a better sense of reality. For most situations, fortunately, the force needed to pull something across a surface mainly tracks the force (gravity, for instance) pushing that something against the surface. For any pair of surfaces, the relationship between those two forces—force needed to pull it along the surface and the weight of the object—defines the friction coefficient. In slightly different words, the higher the value, the more force it takes to move an object across a surface relative to the force pressing it into the surface.[5] Anyone who has pulled a varying number of youngsters on a sled on level ground will be familiar with the relationship between load and required force. Figure 3.2 gives the physical setup one should bear in mind, with perhaps a bit more formality than we'll actually need here.

So what values of friction coefficient do we encounter? For the present purposes we want sliding friction (or kinetic friction), the kind that opposes ongoing motion, as opposed to static friction, the kind that opposes getting motion started in the first place. (Bear in mind that a high friction coefficient is bad if sliding along cheaply is the game.) For sliding friction, values of wood against wood are hard to find since we no longer use wood as both halves of a pair of sliding, bearing surfaces. For wood against metal, specifically oak against cast iron, I found a value of 0.49 if dry, or 0.075 if greased, more than six times less. (As a simple ratio, friction coefficient has no units.) Aluminum against aluminum gives an especially high value of about 1.4. Values for steel against steel are around 0.5, much better, and using hard steel and grease can reduce that value by tenfold. If

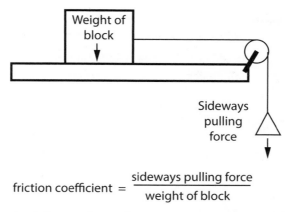

$$\text{friction coefficient} = \frac{\text{sideways pulling force}}{\text{weight of block}}$$

Figure 3.2. The definitional setup for determining friction coefficients. The sideways pulling force here is supplied by a weight on a rope; a pulley converts the weight to a lateral force. In practice, the weight would be replaced by a spring scale, force transducer, or a bucket to which water was added until the block began to slide. As you can imagine, the rig is much more practical for static than for sliding friction!

you have to do without a lubricant, some (but not all!) of the plastics give spectacularly low values. For instance, Teflon against steel has a value of about 0.04. There's no lower limit (in the real world only positive values, of course, can occur), and our joints, lubricated with so-called synovial fluid, put the previous values to shame—values can be as low as 0.001.[6] If you ever debone a leg of lamb, try flexing either the knee or the hip before consigning the bones to the stock pot. You'll feel an amazingly friction-free joint. Not that they're better than our best lubricated metallic joint—in the end, lubrication of whatever origin and however applied seems to hit about the same practical limit.

Oh, yes, an exception exists among the woods, one alluded to in the previous chapter. Lignum vitae (also called "holywood") offers not just unusual density and hardness but also a peculiarly self-lubricating behavior that comes from its high content (about 30 percent) of waxy resin. It absorbs little water and, unlike metals, laughs off the corrosive effects of seawater. So it has enjoyed (if one can say that for an application that's fatal to the tree) long use as the bearing surface for the final driveshafts of ships and for the turbines of hydropower plants. The service life of bearings in which metal shafts turn in lignum vitae housings can approach a century. Sliding against steel, lignum vitae does as well as nylon and nearly as well as Teflon.[7] The parent tree grows slowly, and all too much has already been harvested, but the wood can still be obtained commercially—see, for instance, lignum-vitae-bearings.com. Lignum vitae provided the material for many of the larger (hence non-jewel) bearings of Harrison's marine chronometers.[8]

Allusion to the use of animal fat for old wagon axles, to greasy surfaces, and to lignum vitae brings up the general subject of lubrication. As the old saw goes, "It's easy if you're greasy," although I believe the adage alludes to something quite different. Grease may be good, but the ordinary biological greases—mainly animal fats but also some heavy lipid extracts from seeds—give far from ideal performances. Friction, we remind ourselves, liberates heat. So, even with generous application, greases soften and may melt completely as they warm up. And therefore they tend to be squeezed out of mobile joints at just the worst times. Until a few decades ago, before adoption of sealed, permanently lubricated suspension and steering joints, one took one's car in for "grease and oil" rather than a mere oil change. Oils become less viscous as they warm up, with the same result—hence the role and concern of oilers tending old machinery. And hence the impetus for development of modern multi-viscosity motor oils. The name is something of a misnomer, since an oil that behaves as a 10-weight when cold and a 30- or 40-weight when hot in reality merely avoids most of the viscosity drop that normally accompanies a rise in temperature.

We've devised endless techniques to make sure the lubricant gets (and stays) where it's needed. Grease is packed within the housing of the speed reduction gears on some small electric motors; the heat generated as the motor runs softens it, helping it ooze between bearing surfaces. Grease seals surround packed bearings on many a shaft. On the ends of some electric motors you can see the tiny hinged caps atop the oiling tubes; the tubes lead the oil you periodically add (or should add) down to the shaft bearings. The outer surface of a bearing may be made of bronze that has been fused from powder—"sintered bronze" or "Oilite" bearings—which retains the oil. Enough oil to last many years can be held in a thick felt pad surrounding such a porous bearing. The crankshaft of an automobile, turning in the pool of oil in the crankcase, can usefully splash oil up into the cylinders, as in some early automobiles. Or oil can be pumped into bearings and then returned to the crankcase through grooves cut into one or the other bearing surfaces, as in modern automobiles. And on and on.[9]

Rolling usually meets less resistance than does sliding. Thus rough logs, a little oddly, are often better than smooth rollers to ease the way when pulling something over a rough surface. That difference underlies the advantage of ball bearings, bearings in which a ring of hard metal balls runs around between shaft and housing. Some balls are loose in their races—I once found this out the hard way when disassembling the pedal mechanism of a bicycle. Some are positioned in a light cage, and some of these have spacing devices to keep them properly distributed circumferentially. As you can imagine, a little sliding will always remain, either between adjacent balls or between balls and housing or spacing device. But it can be kept low and, as a significant benefit, the sliding force needn't increase much (if at all) as the load on the whole thing increases. A disadvantage of ball bearings is that loads are carried at points, the contact points between shaft and balls and between balls and housing. Now, a force acting on a point implies infinite stress, the force per unit area, which is a critical factor determining what a material can withstand. Given the reluctance of the real world to handle an infinite level of anything, in reality something will give a little—nothing is perfectly rigid. That deformation generates heat, produces wear, and raises issues of loosening under load. Also, although perhaps not so important, the components of a ball bearing assembly have to be made to a high level of precision and must retain a good fit over a long service life.

A partial fix for all these problems appeared at the end of the nineteenth century, a solution initially intended for wagon wheels. Cylindrical rollers can replace balls and perform the same function. Doing so they have more real-world contact area between the parts of a bearing assembly since contact happens along a line, not at a point. But even better than cylindrical bearings are ones with rollers tapered into long cones—and, of necessity, rollers on axes no longer quite parallel to that of the main axle. Cones enjoy one huge advantage over cylinders, which underlies their use in countless applications whose common feature we rarely notice. Once their tapers are fixed, their fit can be adjusted just by pushing them axially. That may sound all too subtle, but just think of the conical sides of nesting plastic cups and contrast them with the typically cylindrical sides of glass tumblers. Less precision in manufacture may be needed for conical than for cylindrical components, especially rollers. And compensation for wear or alteration of snugness of fit takes only a small axial adjustment. Drill holders in lathes secure the drill with no fasteners—pushing the drill into what's being drilled pushes the conical taper of the chuck ever harder into its housing. Modern automobiles put tapered roller bearings in the joints subject to the greatest abuse, those between the axles and the wheels, as in figure 3.3.

One more piece of the bearing story will head us back to potter's wheels, as promised at the start of the chapter. Thrust-resisting bearings in particular, as the ones below rotating potter's wheels, need a few more words. On the one hand, as already mentioned, nothing is simpler than a sharpened point on either the rotating shaft or the fixed base into which it fits. Either way, the sharp tip of one surface sits in some kind of indentation in the other, perhaps in a wide, conical dimple. On the other hand, tips do not stay sharp, and for even a modest downward force, the stress on the contacts will be enormous. Wood provides the poorest of thrust bearings; stone is better, but it

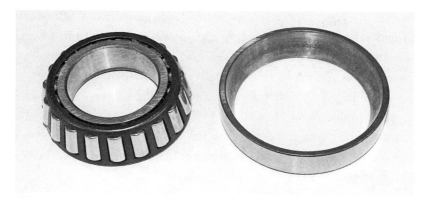

Figure 3.3. Tapered roller bearings and their raceway for a front wheel of a middle-aged rear-drive car. The conical taper and raceway allows the bearing to resist both radial and axial loads, and it also allows easy adjustment of the snugness of fit of bearing assembly and raceway. We trust our lives to such unseen items.

suffers from the peculiar brittleness of all reasonably hard stones, at least by comparison with metals. Hard minerals such as the rubies used by watchmakers give good service, at least on a small scale. Low-friction plastics work, but they do so only if the system tolerates a reasonably broad contact area—none are especially hard. We mainly make thrust bearings of metals. The simplest discrete thrust bearing I've seen is a single bearing ball of hardened steel loose in the well of the socket that accepted the spindle of a turntable for vinyl records. "Loose" meant that I almost lost it when exorcising my youthful unhappiness at owning anything whose inner workings remained unexplored.

In wide use are ball bearings, either loose or in assemblies, which appear in items as humble as the lazy susan on our breakfast table and the swivel of the desk chair in which I now sit. They, too, will spontaneously distribute themselves across the substratum when such things are taken apart. As with sleeve bearings, more severe service calls for tapered roller bearings, either on the end of the rotating shaft or, perhaps more often, on the housing into which the shaft is inserted. Tapered roller bearings work especially well where loaded as both sleeve and thrust bearings. Those on the front wheels of your car face just that combined loading regime when you take a corner or, worse, graze a curb.

These pages only slip along the edge of the subject. A look at a proper handbook for the relevant engineering, such as any of the dozens of editions (an author envies its longevity) of *Machinery's Handbook*[10] will leave you (1) dazed and (2) impressed with the sophistication of contemporary design work.

Back at last to potter's wheels. Or almost so—we need a few words about what they accomplish and how they're operated to put them in some functional context. With a modern potter's wheel (fig. 3.4), the potter first puts a properly prepared lump of clay in the center of the disk. As the wheel is turned, light pressure from a fixed hand or tool forces it to the center. Pushing down with one and then several fingers in the middle of the clay makes an initial hole; further manual pressure with wet fingers and hand enlarges the hole and thins the walls. After final shaping, the wet pot is cut from the wheel and set out to dry prior to firing. The whole process asks for a combination of technical skill, artistic ability, and manual dexterity.

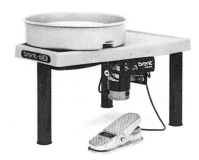

Figure 3.4. A modern hobbyist's electric potter's wheel, with a foot-operated speed control. The turntable weighs about 15 pounds (with the more rapidly turning motor adding further momentum); the entire appliance weighs 130 pounds (Amaco Brent Model EX; www.amaco.com).

The rapid turning of the wheel, typically two to four revolutions per second, minimizes the movement required of the hands. Light pressure on a fairly rapidly moving piece of clay turns out to be more satisfactory than heavier pressure applied to a slowly rotating piece—or at least it reduces any tendency to push outward too strongly and either crack the wet, plastic clay or to produce something unshapely and uneven in thickness.

The operation demands a heavy rotor, in particular one that, when turned, develops a lot of angular (as opposed to ordinary linear) momentum.[11] Otherwise the revolving, evolving pot will slow down too easily when pressed on by the potter. Angular momentum depends on three factors: the rate of rotation, the mass of the wheel, and, especially, how far that mass is centered (as a ring) from the axis of rotation. So the rotor needs not only to be heavy, but needs to have most of its mass quite far outward from its rotational axis.

One commonly encounters some allusion to a role for high angular momentum (or rotational kinetic energy) in assisting the potter by supposedly drawing the wall of the developing pot outward.[12] The effect could not be more familiar—just turn around while holding a weight on a string extending from one hand, and watch and feel the weight pull outward. But the particular aid to the potter of angular momentum seems unlikely for pots of reasonable size turning at reasonable rates. Calculating the outward force on a given weight of clay is a simple matter—or, in reality (for purists), calculating the force that a radial string would have to exert to prevent the clay from being flung outward. The main datum we have to put into the calculation is the requirement that the speed at which the pot's wall passes across potter's fingers should be greater than about 2.5 feet per second (0.75 meters per second).[13] As an example, consider a nice big pot with a wall 8 inches (0.2 meters) in diameter. If the wheel turns twice each second, that wall speed will be 4.2 feet per second (1.25 meters per second), easily fast enough to allow hand forming. At the same time, the effective outward force on a bit of clay will be only 1.6 times the gravitational force, the downward force of the weight of that same bit. Adjusting the assumptions about pot diameter makes little practical difference to the result. So by turning, the potter's wheel gives only the most modest of outward pulls; the outward pull only slightly exceeds the downward sag from gravity, to which the clay ought to be quite indifferent if properly prepared. Put another way, if the clay is soft enough to be pulled appreciably outward, it won't retain its shape but will sag downward under gravity too easily to be worked.

Mass (as weight) would present no problem to stone-based cultures making potter's wheels, but spinning that mass in the form of a wide disk would put a heavy load on whatever serves as a thrust bearing at the bottom. The poorer the bearing, the more the friction it generates, and the greater the force (technically, the moment or torque) that's needed to turn the wheel. Unlike pots, potter's wheels are rarely if ever preserved, but clever archaeologists can figure out quite a lot from detailed analyses of the pots themselves. In short, slow turning is relatively easy and was common, while fairly rapid spinning, as needed to "throw" a pot (even without much centrifugal contribution), presents considerable difficulty and has been rare.

Not surprisingly, the first turning devices apparently just allowed the work to be rotated while the potter added clay. The potter most likely added clay as long, cylindrical worms winding in a coil around the upper edge, and in adding it the potter did much of the shaping. A proper thrust bearing wasn't truly necessary, although some centering arrangement short of a fully weight-supporting bearing would have been a convenience. Strictly speaking, such coil-formed pots were not wheel-shaped, much less wheel-thrown. But wheel-shaping can help, and it appeared quite early. Forming a pot on a rotating, pivoted turntable allows bumps and depressions to be located, either by feel or with the aid of an externally mounted stick or template. That produces a more symmetrical product. For a given strength, such a pot would have, on average, thinner walls. Thus it would require less clay to begin with and would weigh less in the end. In many parts of the world, pots even now are commonly formed on turntables that revolve slowly. Even New World potters, who never turned to wheel-thrown pottery, made use of turntables.[14]

Ancient images show wheels supported and turned by a pair of helpers. Some version of such wheels came into use in the Indus Valley of India, in Mesopotamia, and in Egypt well before 3000 BCE, and in China at some comparable point in antiquity. As noted, the earliest ones would have been turned slowly, not spun, both as a logical developmental sequence and by the evidence of their narrowness, as in figure 3.5. As explained earlier, spinning works poorly unless a wide flywheel confers enough angular momentum—again not to fling clay outward but to produce smooth rotational motion despite the friction of molding hands against clay. Real flywheels, known in Egypt from about 3000 BCE, had to have a heavily loaded thrust bearing at the bottom. One might expect that a supporting ring or sleeve bearing would be needed to keep the wheel from wobbling, but that's not strictly necessary. A wide, heavy disk with a reasonable bearing

Figure 3.5. An ancient Egyptian potter's wheel. The two operators employ both hands and feet in turning it. The person on the right seems to be kneading clay with his feet (from a tomb at Thebes, about 1450 BCE, via Scott, 1954, p. 389).

Figure 3.6. An operating kick wheel, this one of a deliberately anachronistic demonstration design, at Museum Village, Monroe, New York.

in the center of its bottom will turn, much like a child's toy top, with sufficient stability to remain horizontal—especially if turned by a pair of experienced workers with no direct hand in forming the pot. Minimal clearance above or below the supporting surface will prevent serious tipping if the thing gets out of kilter. Unless the potter's wheel has a kick wheel, instead of helpers, as in contemporary non-motorized wheels (fig. 3.6), that center thrust bearing should be sufficient. With a kick wheel, the lower wheel is the heavier of the two, which keeps the center of gravity low and reduces the load on the now-necessary sleeve bearings, one typically located just beneath the working disk and the other beneath the kick wheel.

Of what were these early thrust bearings made, and how were they shaped? V. Gordon Childe gives a good illustrated discussion, and recent work has added many more examples—pots were important yet fragile enough so the potter stayed busy, and stone artifacts persist.[15] Sometimes the wheel's bottom had the convexity and the fixed base had the socket; sometimes the wheel had a concavity in the bottom and the base had some dome-shaped convexity or tenon. Childe even cites examples of intermediate bearing stones, with tenons extending both downward, to fit a socket in the base, and upward, to engage a hole in a wooden kick wheel.

Questions such as the one with which the chapter began occur even to the casual inquirer. Was the wheel and axle of the cart ancestral to the potter's wheel in the sense of offering some model of what might be possible? Or was the potter's wheel ancestral in the same sense to the locomotory wheel and axle? Or were they independent innovations that performed their tasks in insufficiently analogous contexts? As far as I can see, the technologies themselves offer few clues. A potter's wheel centered and turned by assistants seems an unlikely model, but so does a crude, horizontally rotating cart wheel. "Unlikely" certainly does not mean "impossible," and the temporal and geographical coincidences do suggest some relationship. That is, unless they indicate nothing more than the time and place of a generally innovative technology, which seems reasonable as well. Certainly pottery making originated far, far before wheel turning, probably as far back as 20,000 years BP.[16] But for obvious reasons, nomadic cultures didn't produce pots in profusion, so the record, one might say, is fragmentary.

An equally curious question comes from the different styles of pot making in the Old and New World. Throwing pots on potter's wheels is a Eurasian technique, unknown in the Americas even when Europeans arrived, about five thousand years after the appearance of the devices in the Middle East. And American pottery was anything but unsophisticated. Pre-Columbian sites have yielded great treasures of fired clay creativity, with no scarcity of well-rounded vessels. But pots continued to be built up by coiling long worm-like cylinders of clay, if with occasional use of turntables for finishing. Pottery was ubiquitous in the Americas from very early times. Does this imply that without wheeled carts as models, wheel-thrown pottery does not develop?

And that once again brings up the question of independent origin versus cultural diffusion in Eurasia. Put more specifically, did the potter's wheel have a single origin, or did the wheel start separately in, for instance, China? After all, the Chinese discovered porcelain, arguably the finest ancient ceramic, as much as a thousand years before its appearance in the West.[17]

Perhaps a book about turning things ought occasionally to step back and reexamine the case for turning in the first place. We take it for granted that a spherical, cylindrical, ellipsoidal, or other shape characterized by a round cross section is the default for functional ceramic objects. Such objects do not pack as well as ones with rectangular or hexagonal cross sections, and they're all too prone to rolling around in a way hazardous to the integrity of such brittle artifacts. For pots in particular, bottoms are most often narrower than bellies, which reduces the force it takes to tip one over. Why not make flat-sided pots, or, if they have to be round in section, why not cylindrical or even conical pots with enlarged bases?

We do run into slab-sided pottery of all degrees of antiquity, but it rarely comes in the form of vessels for storing supplies or for carrying loads. Making it is easy enough, since clay works as well for gluing sides together as it does for making those slab sides in the first place. I once saw a large, rectangular ceramic coffin in a Middle Eastern museum. Building slab-sided artifacts may be simple, but the products have not one but two peculiar and interrelated disabilities. A hint of the problem appears in the way our air-handling ductwork may be rectangular in cross section, but our pipes for liquids are always round. The difficulty comes not so much from the way a rectangle might be prone to fold as it is with, first, flat walls and, second, sharp corners.

Consider an ordinary milk or juice carton, as in figure 3.7. You can tell whether it's full or empty at a glance—just look at whether the walls bulge. Where a wall feels a pressure difference across it, the wall has a strong tendency to become rounded. The concave side faces the higher pressure, the milk, juice, or, less conspicuously, the wine within its liner in a cardboard carton. And the greater the

Figure 3.7. Two soft-sided containers of rectangular cross section. The bulging sides leave little doubt as to which one is full, not empty. I've drawn a pair of straight lines (marked with arrows) to emphasize the difference.

curvature, the better the wall resists the internal pressure.[18] The greater curvature of their narrow walls is why the skinny and flimsy-looking tires of racing bicycles withstand far higher pressures than the fat tires of heavy construction vehicles. It's why, as a trip to the building supply store will show, wider plastic pipe of the same pressure rating (Schedule 40, perhaps) has thicker walls. Worst of all for withstanding transmural pressure is a flat wall—zero curvature means zero resistance to pressure differences across it. What makes flat walls at least minimally functional is that their thickness in effect provides some effective curvature—a curve concealed within the wall. The only flat wall possible on any inflated balloon is one pressed against a similar wall that balances the pressure. A circular cross section best distributes the stress on a wall, and so it gives the most resistance to cracking or bursting relative to the amount of clay needed to make a vessel—as well as the least non-useful weight that needs to be transported. If made of rubber of uniform thickness, an inflated balloon will always have a circular cross section, whether it's spherical, ellipsoidal, cylindrical, or of periodically varying diameter. That's its way of equalizing the stress on the wall—any place where the stress is greater will automatically stretch further and restore the circular cross section.

Sharp corners are locations of particularly high stress. You nick some fabric or foil to get a tear started. Modern jet aircraft have windows with rounded corners, something learned the hard way with several catastrophic failures of the first, the de Havilland Comet, in the early 1950s. The gaps between the teeth of modern gears have rounded bottoms. For most materials, rounding yields a remarkable increase in working strength. On a car I once owned, a door handle broke; I noticed that the side normally hidden by the door had sharp internal corners. On the manufacturer's replacement, those corners were rounded—someone had wised up. So slab-sided vessels develop extra stresses in their walls and then concentrate the resulting forces on their inside corners. This clear cause of functional fragility can be alleviated only with extra material in the corners, which means waste and weight.

One traditional culture did make containers that were rectangular rather than circular in cross section. These were the Indians of the northern Pacific coast of North America. Their vessels, though, were made of closely fitted (and relatively thick) slabs of wood, slabs split from the trunks of the red cedar trees critical to so many of their ingenious contrivances. They even cooked in the wooden containers, heating rocks and periodically immersing them in these strange stew pots. Wood is remarkably good stuff if you have the requisite tools and craftsmanship. It's tough material, with far less of the brittle fragility of pottery; it's less dense than ceramics, so thicker walls weigh no more; and it obligingly swells when exposed to ordinary liquids— our best wine and whiskey barrels remain wooden. Neighboring

Indians made plenty of pottery, but these coastal people, the most sophisticated hunter-gatherer culture anywhere ever, did not judge the art worth the trouble.

What about the bottoms and the overall width of vessels? A flattened bottom raises the same problems as flat walls—the problems of reduced pressure resistance and of force concentration at the bottom-wall junction. The smaller the bottom, the less the problem, so the practice of tapering a vessel down to a narrow bottom has some justification despite the greater tippiness of the product. Ancient vessels often had rounded bottoms that formed portions of spheres and consisted of simple continuations of the walls. Keeping these round-bottomed flasks upright asks for some indentation in whatever surface supports them. On the one hand, that's a nuisance, but, on the other hand, it's a handy way to ensure stability even when the vessel isn't exactly upright. The Greek amphoras that we recover in vast quantities from sunken freighters in the Mediterranean usually had bottoms that taper to a blunt point. That facilitated insertion into racks in the holds of what must have been infamously rolling ships. The insides of the amphoras, though, had properly hemispherical bottoms. These containers, by the way, are not the wonderful amphoras of art museums but cheaply made terra-cotta vessels that were rarely reused—near Rome a wide hill, about 150 feet high, consists mainly of the accumulated fragments of broken amphoras. Width? Narrower is stronger, relative to either pressure within or external crushing. So the spherical shape that gives the greatest volume relative to wall area is no longer the functional optimum in terms of volume of clay consumed in fabrication.

Then there remains the issue with which the book began. How was the wheel, here a potter's wheel, turned? Neither wind power nor waterpower, both usually designed with rotational prime movers (windmills and waterwheels), found much use turning potter's wheels. Early potter's wheels depended on one or two people helping the potter, repeatedly grasping the outside of the wheel, turning it, and then releasing the grasp to grasp again farther around. Sometimes wheels appear to have been fitted with peripheral pegs for temporary handholds. For a very long time, the most common potter-operated arrangement has been the kick wheel. It works basically the same way—the potter puts a foot on the wheel, turns it, in an ergonomically awkward lateral movement, takes foot off wheel, moves foot back again, and repeats the cycle. A kick wheel, of course, constitutes a second rotor that must be located several feet below the pot-bearing wheel, so sleeve bearings, as mentioned earlier, become unavoidable. At the same time, the kick wheel itself can provide most of the angular momentum, so the upper wheel has fewer mechanical constraints, such as needing a wide, heavy periphery. This turning mechanism—repeated cycles of grasp and release—has been so widespread that it merits a chapter of its own, here chapter 6.

[4]

Going in Circles

Could there be a simpler way to make animal muscle drive fully rotating machinery? Just make the animal walk around and around in monotonous circles while harnessed to, say, an arm extending outward from a capstan, as in figure 4.1. The whole engine turns, neatly circumventing any awkwardness of making muscle, that linear engine, produce continuous rotation. Until practically yesterday—that is, until about a century ago—animal-powered devices of this kind performed a wide range of tasks nearly everywhere in the world, and even today they can occasionally be seen in rural third-world areas. Still, the technology draws little attention. In its heyday (or hay day, perhaps), it was mostly too ordinary to attract notice; in the so-called living museums of yesteryear, working models don't abound. Horse wagons and oxcarts, perhaps, but these stationary devices applied their engines in ways all too likely to offend contemporary sensitivities.

Harnessed in this way as parts of rotary mills, animal power enjoys several advantages. Since the input consists of hay, grain, and the like, no fossil fuel or electricity need be purchased. The effluent manure makes good fertilizer. Cultural taboos permitting, worn-out engines provide excellent human nourishment. Better, a nice

Figure 4.1. An ox works a chain-operated well pump in China; one of a large collection of photographs of irrigation devices made available by Thorkild Schiöler, at www.kattler.dk/schiolers/uk/index.html. This image can be found at http://www.kattler.dk/schiolers/uk/160.html.

flexibility comes with the territory. If a horse pulls a wagon, operation is maximally efficient at fairly specific values of speed and load. Thus asking a given wagon to travel on a given kind of road limits the immediate options. If an animal turns a revolving capstan, the designer has access to a wide range of speed-load combinations simply by changing the length of the radial arm, that is, by changing the radius of the circle in which the animal walks. Still better is the versatility provided by choosing the size and number of animals that worked simultaneously, from a single dog (or child) to more than a dozen horses. In the nineteenth century, horses working remarkably simple winching arrangements pulled entire houses—on rollers and slowly, perhaps, but at times even going up slopes.

While these turning mills have attracted less interest than either moving vehicles or potter's wheels, they represented serious power sources for at least three millennia. Again, they performed a huge range of tasks—grinding grain; pumping water; lifting loads from vertical mine shafts; sometimes winching boats, plows, and large logs; and providing prime movers for almost every heavy task in pre-industrial factories. A wide diversity of what can best be lumped under the term "transmissions" coupled animals to tasks, sometimes in fairly complex ways.

Just about every domesticated equid or bovid served in one instance or another as an engine—oxen, yaks, water buffaloes, donkeys, mules, and horses—plus camels, dogs, goats, and humans. And possibly elephants, reindeer, and llamas. Bears, semi-bipedal, sometimes turned treadwheels in medieval Europe. Naturally, a minimally useful animal size sets one limitation, especially since mills were almost always turned by animals that had been previously domesticated for other purposes than this power-providing business. If you could harness it to cart or plow or make it carry a pack, then you could make it turn a mill. Chickens and rabbits remained purely sources of meat.

We ought to distinguish among several versions of these animal-powered rotational devices. No surprise—the names given them are inconsistent and sometimes even illogical. So I'll assert the author's prerogative and set up a scheme, sticking with names that seem reasonably ordinary and descriptive. The sequence follows what seems a likely historical progression of time of introduction, a sequence based on technological difficulty.

- *Direct drive* (fig. 4.2a). The animal or animals turn some radial extension of whatever basic device needs turning. No gearing or other linkage enters the picture, although altering the length of the arm provides a way to change the turning force and rotation rate produced by an animal or team that walks at a given speed.
- *Capstan* (fig. 4.2b). Here the radial arm again extends outward from a drum or capstan turning on a vertical axis. Now a rope or wire from the capstan pulls on some load adjacent to but

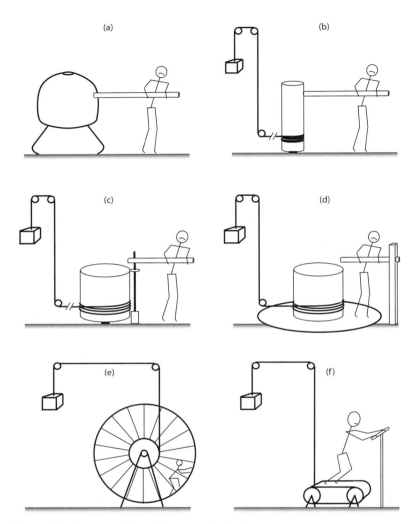

Figure 4.2. Six ways of coupling an animal—here a human—to a rotational transmission: (a) directly turning some mill; (b) circling a capstan; (c) circling a capstan with intervening gearing; (d) walking in place with the capstan on a rotating disk; (e) climbing a treadwheel (here inside); and (f) walking a treadmill. The hash marks on the rope in drawings (b) and (c) indicate an indefinite length of rope; the load may be quite remote from the capstan.

not ordinarily touching the drum. Or, with some equivalent of what we know as a pair of bevel gears to turn a right angle, a rotating shaft may extend outward from the capstan beyond the circle in which the animal or animals turn.

- *Internal gearing* (fig. 4.2c). A rope or cable extending outward from a drum of modest size provides, as we'll see, a huge force together with very slow motion. Many tasks do better with the opposite: smaller force and faster motion. The engine allows little adjustment, perhaps a bigger or smaller animal or more or fewer of them. But all manner of linkages—transmissions— within the drum can trade force and speed against the other.
- *Treadwheel* (fig. 4.2d). This changes the frame of reference, much as Copernicus changed the frame of reference for our solar system from the Ptolemaic earth-centered outlook to a much simpler sun-centered (heliocentric) view. Instead of asking the animal to move around on a fixed substratum, now the animal must walk in place atop a disk that keeps revolving as a result of the animal's efforts. A drum on the rotating disk then drives various devices farther along.
- *Wheel treadmill* (fig. 4.2e). Here again the animal walks in place, but now it walks on either the outside or the inside of some large, rough-surfaced cylinder whose axis is horizontal. It needn't move the legs on one side of its body farther than legs on the other or contend with the curved path enforced by all but impractically large disks. Offsetting these advantages, it no longer walks on a level surface.
- *Belt treadmill* (fig. 4.2f). Walking on the surface of some form of moving belt retains the advantage of walking straight and has the additional advantages of not walking on a surface of variable slope and of avoiding climbing—or else allows climbing at a predetermined and usually adjustable slope.

As noted in chapter 2, wheeled vehicles drawn by animals appear between 3000 and 4000 BCE. Applying the same animal traction to the equivalent task of turning a vertical shaft does not make an appearance for perhaps another 2,000 years, to me a surprisingly long time. In the interim, oxcarts, horse chariots, and the like had become anything but novel, taking their place as standard technological fixtures of much of the Old World. A device in which one or two animals turn a vertical shaft should be more forgiving as far as shaft strength and bearing friction than one in which wheels and axle are pulled forward, loaded downward, and bumped irregularly.

So I think that temporal gap deserves examination. My best explanatory guess is basically a utilitarian argument—what, exactly, would the direct drive device (fig. 4.2a, again) accomplish; what would it displace or augment or speed up? Large-scale rotary motion must not have counted for much in those innocent times, with no rotational, vertical-shaft water pumps and certainly with no rotational posthole diggers. They may have used drills and augers, perhaps, but the small size of these tools doesn't merit the employment of one or more large creatures that produce high levels of power when turning capstans. At the same time, the very idea of rotational machinery would have represented a degree of novelty quite unimaginable for anyone whose entire experience revolves around ubiquitous electrical supplies and fractional horsepower rotating electric motors within appliances.

Yes, lifting water mattered to early cultures of the Mediterranean, Middle East, and India. Rivers, reservoirs, and shallow wells might provide the wherewithal for irrigation, but all too often farmers could not rely on gravitational delivery. A case has been made and generally supported that the need for irrigation provided the impetus for social and thus political organization, explaining the origin in the Middle East of the earliest villages. The most common irrigation device, beyond bags or buckets that were lifted and carried, consisted of something most often called a "swape" or a "shaduf," a pivoted, counterweighted pole, with no rotation except the slight arc of its swing, as in figure 4.3. A swape improved things in two ways. First, it allowed the person to pull downward instead of lifting upward, an

Figure 4.3. A swape, this one as used by the Anglo-Saxons. Visible at the top is the bulbous counterweight; the other end of the lever is attached to the rope that extends downward into the well. (From Ewbank, 1842, working from an old manuscript in the British Museum; he notes that, in reality, the bucket's lever arm was probably much longer.)

easier repetitive movement for a human. Pulling down represents, in fact, a fairly efficient use of human muscle, as the first measurements of human-power output in the eighteenth century determined. And second, the relative lengths of pole on the two sides of the pivot permitted adjustment for the depth of the water, trading lift distance against lift load per cycle. The devices, under a confusing host of names, saw both long and widespread use.

Pumping arrangements depending on continuous rotation, in particular bucket chains, came much later. A bucket chain, for instance, requires that some kind of gearing convert the vertical shaft rotated by a turning animal into the horizontal shaft at the top of the chain. That horizontal shaft might instead be turned with a crank, but these were quite rare until the last thousand or so years. A less handy alternative, handspikes, do not seem to have served this particular purpose. We will meet one device, a peculiar treadmill, for doing this turning continuously in the next chapter.

Grain-grinding mills were probably the earliest direct-drive rotational devices. Domestication of grains formed central components of the agricultural revolutions of both Eurasia and the Americas—rice, wheat, barley, oats, corn, millet, and others now of lesser or no importance. They provided—and still provide—most of the components necessary for a human diet and, after drying, they can be ground and then stored compactly for later consumption with minimal spoilage. With just the addition of a small amount of any legume, a human can get along without eating much else than one or another grain. Reconstituting dried grain takes little more than soaking, and every kind of grain can be boiled to produce a palatable porridge—especially when flavored with locally available herbs.

But we humans have a long tradition of culinary experimentation. Dry grain can be ground into meal or flour of varying degrees of coarseness, mixed with water and various flavorings, and baked, grilled, or dry-fried into a simple flatbread. Left to ferment for a while, the wet batter can be baked into leavened flatbreads or loaves. So food preparers have for millennia devoted considerable time and energy to grinding grain—and, incidentally, grinding smaller quantities of a wide variety of other materials. Our literature abounds in metaphors alluding to grinding grain. That's what Samson did in the Philistines' prison house, a dull job is a "grind," and the hack writer "grinds" out prose.

So how to grind grain? The ancient housewife's tool—it seems to have been women's work for the most part—was a quern, basically two stones rubbing together with the kernels of grain sheared between them. Mortar and pestles may share similar antiquity and

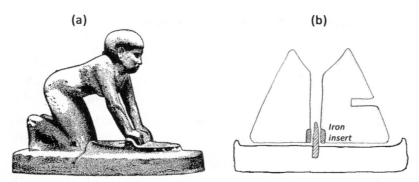

Figure 4.4. Querns. (a) A saddle-quern, Egypt, ~2500 BCE. (There is a similar photograph at http://teachmiddleeast.lib.uchicago.edu/historical -perspectives/the-question-of-identity/before-islam-egypt/image-resource -bank/image-13.html.); (b) beehive-quern in cross section, with centering pin and iron-bearing ring (from http://goblin-online.net/genealogy/bedgreave /bedgreave_history.html).

basic mechanism, but they're more appropriate for small batches of relatively valuable items—spices or soft minerals—than for useful quantities of grain. Being stone, querns persist for archaeological scrutiny, so we have no shortage of examples. In addition, their use was nearly universal in geographical scope and extended over millennia, beginning long before we took up agriculture. We've ground wild grain, pigments, and much else, for, at the very least, the past 50,000 years.[1] In early (and many later) querns—called "saddle querns" or, in the New World, "metates"[2]—the top stone was moved back and forth in a saddle-shaped depression in the lower one (fig. 4.4a). Basalt was the preferred material—it stayed rough enough to keep doing the job while shedding little grit into the resulting meal. But many different stones served in one place or another. By 6000 BCE, a coastal village in Turkey was producing saddle querns in commercially exportable quantities.[3]

While some querns (but none of the really early ones) were designed for a circular motion, the action most often must have consisted of an arc in one direction followed by an arc in the other—

convenient for a person operating one cycle after cycle without having to release her grip. To be at all convenient, full, repeated rotation demands a crank to avoid that awkward repositioning of the hands; and, again, we have scant evidence of cranks in antiquity. What evidence we have is ambiguous, a vertical, peripheral hole in the top stone, one into which a wooden handle could have fitted. We'll get back to cranks in chapter 8. A more common arrangement had horizontal holes in the top stone into which wooden extensions must have fitted. These would have been convenient for a back-and-forth movement either by the same person who poured the grain in at the top or by an assistant. The most common version has been called the "beehive quern" for its overall form (fig. 4.4b). A typical one in the British Isles before and during the Roman occupation was about a foot in diameter.[4]

Only when grain grinding became industrialized did direct-drive animal-powered machinery come into use. "Industrialized" in practice meant commercialization, not simply of dry grain, but of ground grain, offered as a product to city dwellers, bakers, and the like. The mercantile Greeks and Romans kept the basic rotatable quern, now evolved into a more conical shape. The Romans called it a "donkey quern," the name applied, it seems, whether the drivers were donkeys, oxen, humans, or any other animal. Typically, two donkeys, hitched with the inappropriate bovid yokes of the time, turned a single quern, as in figure 4.5. Adding to that inefficiency, the donkeys walked in a circle only slightly wider than the diameter of the quern, judging both from surviving illustrations and from the distance between the adjacent querns of a large bakery preserved in Pompeii. The second animal may have served as much to balance the applied force as to provide extra power. But however inefficient, these querns were good enough—a pair of donkeys or slaves, engines of the minimum available size, may have provided more power than needed. Roman armies traveled with their querns. Since the quartermasters allowed about one quern per five soldiers, these were more likely to have been small beehive querns than the larger donkey querns.

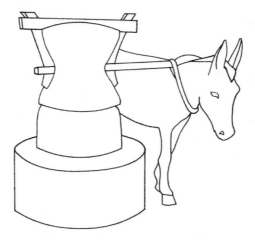

Figure 4.5. A Roman-style donkey-quern, here shown with a single donkey harnessed in the ancient bovid-style yoke (from Vogel, 2001). More commonly, two donkeys would be harnessed symmetrically to the donkey-quern.

One hears little about animal-powered rotary grinding machines beyond the heyday of querns. Water-powered and, later, wind-powered grist mills come in for much more attention. But animal-driven turning continued in all cultures in which animal power—with its associated handling, harnessing, and husbandry—were comfortably mundane. Well-worn millstones abound as artifacts where no other power source can be imagined. Farms in the North American Midwest during the late nineteenth century commonly had very large feed mills in which corn could be crushed. Several versions employed a pair of horses turning a sweep with a radius of about ten feet.[5] These farms were (and their successors remain) large—the Homestead Act of 1862 set as a standard size 160 acres, or a square almost half a mile on each side. Their mills may bear a superficial resemblance to donkey querns, but they're made mainly of metal and have internal bits and pieces unimaginable to the inhabitants of classical Roman latifundia.

In addition, most post-classical animal-driven gristmills turn far faster than any walking animal could directly drive. In a sense, a donkey or horse walking in circles constitutes what an engineer might call a severe impedance mismatch. It has the capacity for excessive force but achieves too little speed. Indeed, as implied earlier, just that limited demand for force by ancient querns may have permitted the extensive use of donkeys and horses long before these equids could be effectively harnessed.

One application of direct capstan turning that is less often mentioned enjoyed (for all but the animal) a far longer history than the donkey quern. Beginning, it seems, with the ancient Phoenicians, Mediterranean cultures pulverized newly mined ores with a device that, in its most recent form, was called an "arrastra," from the Spanish word for "drag," *arrastrar*. Very simply, crushed ore lent itself to roasting or smelting far more readily than the large chunks produced by the miners. This marvelously simple machine (fig. 4.6) consisted of a circular trough with a rounded bottom and low side walls in which the ore was placed. Going around and around in the trough was a heavy stone (or several stones) hitched to a revolving beam.

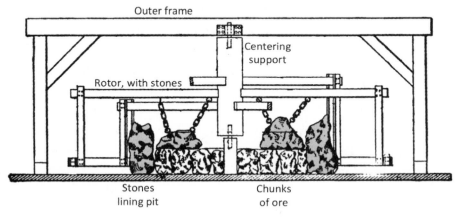

Figure 4.6. A two-stone arrastra in cross section. Crushing stones hang by chains from the rotor. Animals (oxen, mules, etc.) would have turned the rotor, walking in a circle between it and the outer frame (from Richards, 1903, p. 238; simplified by erasure of dimensions, shading, etc.).

The beam, in turn, ran from the central axis of the trough out to the hitching point for the animal that turned it.

Arrastras saw particularly widespread use in Italy and Iberia. From Spain they were introduced into the New World, and some, at least, remained in use up through the middle of the twentieth century in small gold-mining operations in California. The definition, though, is not particular to the direct-drive animal-powered capstans—where available, moving water provided the power. And in their final century, increasing numbers of arrastras were hooked to steam and even internal combustion engines—as described for these relatively recent ones by T. M. Van Bueren.[6]

Again, besides querns, an occasional grain-grinding successor, and arrastras, I know of no other widespread pre-industrial applications of direct-drive animal-powered capstan turning. But what if one adds a simple additional element, a rope extending tangentially outward from the capstan, as in figure 4.2b? Turning the capstan will pull on the rope, and the longer the arm extending outward to the pulling animal (relative to the diameter of the capstan), the more forceful the pull. The capability of this arrangement is worth a look. How much force can a horse-powered capstan exert? As it turns out, quite a remarkable amount, enough to do the heavy pulling and, with a simple pulley inserted, to do the heavy lifting needed to build large buildings, bridges, and the like. Consider a draft horse that weighs about 1,300 pounds (600 kilograms). Well-harnessed and working at its most efficient speed, a little less than two miles per hour (1.7 mph, or 0.75 meters per second, or 2.5 feet per second), it comfortably pulls with a force of about a tenth of its weight, 130 pounds (60 kilograms or 600 newtons).[7] That, incidentally, comes to a power output of 325 foot-pounds per second, about six-tenths of a horsepower as conventionally and somewhat arbitrarily defined.

Now assume that the horse is hitched to an arm extending 10 feet from a capstan 2 feet in diameter, or a foot in radius. A cable from the capstan pulls on an external load, as in figure 4.7a. The arrangement

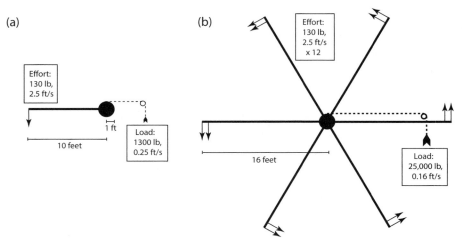

Figure 4.7. Two horse-whims, presented as top-view force diagrams. (a) A single horse whim with a force amplification factor of 10; (b) a twelve-horse whim with a sixteen-fold force amplification.

multiplies the horse's force tenfold, to 1,300 pounds (600 kg, 6,000 newtons). That's enough to lift one end of a small car. What if, as in some large nineteenth-century versions, a dozen horses pulled on six radial spokes at an average radius of 16 feet as in figure 4.7b? That's a pull of 25,000 pounds (11,000 kg or 110,000 newtons). Of course, power isn't amplified—the cable is drawn at only a sixteenth the speed at which the horses walk. Still, for its transmission, that pull would require a manila rope of the best quality fully three inches in diameter or a four-strand wire cable an inch in diameter[8]—serious pulling by any account. By our impatient contemporary standards, horses may have been far from fast, but by turning capstans they could perform the heaviest tasks of industry and construction. Oxen pull at least as well, but they retain nearly their maximum force only up to a still slower speed.

These devices went by a diversity of names, each immediately understandable in its time and place. So we run across "sweeps," "whims," "horse powers," and still others—none with the potency as search terms of, for instance, "querns," but at least better than "mills," which has too many meanings, and "circumambulatory technology," daunting even to a biological scientist. How far back can we trace their use? That's tricky, again because of their sheer mundanity. Absence of evidence is the poorest evidence of absence, so I'm left wondering whether the classical Phoenicians, Carthaginians, Greeks, or Romans used them as winches to haul ships up ramps so they could work on the hulls. We can assert with greater certainty that the ancient Egyptians built the great pyramids without them, however useful they might have been. For that matter, the pyramid builders made no use of animals except as inputs to their communal kitchens. Our confidence comes from the way they chiseled images on stone rather than (or in addition to) recording things in drawings and hieroglyphics on more perishable media. Perhaps wheeled vehicles wouldn't have been up to the necessarily heavy tasks of these monumental public works, but capstan winches turned by oxen would surely have been of enormous assistance[9] and within any reasonable estimate of Egyptian technological capabilities.

Not only does more material survive from the late Middle Ages than from classical antiquity, but technology appears to have come in for more attention from literate and artistic individuals. Mechanical technology also drew the attention of the earliest practitioners of one other and particularly crucial technology—printing. We are, of course, looking at the era just following the wide adoption of the modern mode of printing with screw presses, cast type, and, most importantly, movable and reusable type—the instigations of Johannes Gutenberg in the mid-1400s. A surprising number of well-illustrated treatises appeared in the fifteenth and sixteenth centuries. Among them, the most prominent (in present recognition at least) just happens to have particular relevance here. *De Re Metallica* was published in 1556, written by Georgius Agricola, the Latinized name of the mining engineer and metallurgist Georg Bauer (1494–1555), of Saxony (now part of Germany).[10] *De Re Metallica*'s visibility has undoubtedly benefited not just from the excellence of the English translation of 1912 but from the later careers of the translators, another mining engineer and his classicist wife, Herbert Clark Hoover and Lou Henry Hoover, subsequently president and first lady, respectively, of the United States of America.

So what does Agricola ask of his horse-powered whims (the Hoovers' term, which I'll adopt)? The largest whims pull a rope that extends outward from a drum. Passing over a pulley, as in figure 4.8, the rope hoists a load upward from the vertical shaft of a mine.[11] With

Figure 4.8. Agricola's horse-whim hoist. Note the two-horse whippletree on the beam at the lower left. (A whippletree, of which this is the simplest possible, balances the load on the animals.)

four 12-foot beams, each taking two horses, and a drum 3 feet in diameter, the device would have been capable of lifting about 4 tons of ore—calculating as in the hypothetical example given earlier. In Agricola's figure, two ropes pass over a parallel pair of pulleys. They look as if they wrap around the drum in opposite directions, so one would descend as the other ascended, and the horses appear harnessed on swivels so they could reverse directions. Since the miners at the bottom would have needed some time to fill the containers with ore, the time taken for reversing would not have taxed the overall speed of operation. Nor would the horses have had to turn all that many times—a 200-foot shaft would have taken only about twenty-one revolutions.

By contrast with the vastly more famous (at least to us) Leonardo da Vinci (1452–1519), Agricola produced thoroughly practical devices, ones intended to apply contemporary technology to contemporary tasks—in his case, mainly mining. Moreover, Agricola published; Leonardo squirreled away his notebooks and so contributed next to nothing to the technology of his time. If nothing else, Leonardo set a bad example for our present graduate students, too many of whom never quite manage to reformat and then submit their theses for journal publication.

Not quite as well known to us as Agricola—perhaps because he attracted no such famous translators—is the Italian military engineer Agostino Ramelli (1531–1608). For our present purposes, his elegant compendium, *The Various and Ingenious Machines of Agostino Ramelli*, holds even more interest, the machines being both as various and as ingenious as promised.[12] He offers one device in which a rope extends from a capstan driven by a horse whim. In this one, figure 4.9, the two ends of a rope alternately pull upward and lower a pair of carts that ride on an inclined plane.[13] The task consists of excavating a ditch such as the moat around a fortified town or castle. A single horse provides the power source—nothing so large and elaborate is shown (or needed) as in Agricola's mine hoist.

Agricola's and Ramelli's were neither the first nor the last of this

Figure 4.9. Ramelli's horse-whim inclined lifter. This one both offsets the weight of the carts and minimizes the downtime lost as carts are raised and lowered.

tradition of fifteenth- and sixteenth-century mechanical compendia. They marry three skills that had reached high levels in Europe during that period: practical mechanics, representational art, and (of course) printing. Leonardo da Vinci was only the best known, certainly not unique or even the first individual combining the first two. That award might go to Mariano Taccola (1382–1453), almost a century earlier and a contemporary and acquaintance of the great Florentine dome-builder Filippo Brunelleschi (1377–1446). Taccola both created traditional religious art and published engineering drawings—

although the artistry and attention to detail of the latter don't approach the levels of Agricola, Ramelli, and their contemporaries.[14]

A rope can only pull, which limits the applications of this kind of whim. What's needed for many purposes is some push-pull power takeoff. We might use a pair of pulleys connected by a belt, and a horse-harnessing culture knew how to make good leather belts. The difficulty again is the slow speed of a horse- or ox-turned whim—a very high-force, low-speed engine. To deliver its power, a very strong—meaning a very broad—belt would be needed. The belt would have to be tight (with provision for periodic tightening), which demands a very rigid mounting for both the whim's shaft and for whatever is being driven. A far more forgiving power takeoff consists of making the whim turn a horizontal shaft that extends radially outward from the whim's own vertical shaft. Devices that take this tack abound at least as far back as the time of Taccola, just mentioned, meaning the early fifteenth century. That they did all manner of tasks and drew no particular comment at the time implies a considerably earlier origin.

Extending a radial shaft from a rotary whim creates several difficulties. For one thing, the animals (or people) that go around have to pass either over or under the shaft. An overhead shaft asks a lot in the way of strong and stiff superstructure, while a subsurface shaft requires a surface above it that will bear repeated (and repeated and repeated) treading by the all-too-pointy hooves of draft animals. The overhead version seems most often to have been preferred. In part, that may have been driven by considerations of space—as we'll see, the usual gearing for converting rotation of vertical to horizontal shafts wasn't compact and would have required a considerable subsurface cavity. Beyond that, wooden components don't take kindly to subterranean dampness. Bear in mind here and elsewhere that profligate application of metals characterizes only the past few centuries—with the slight exception of military applications.

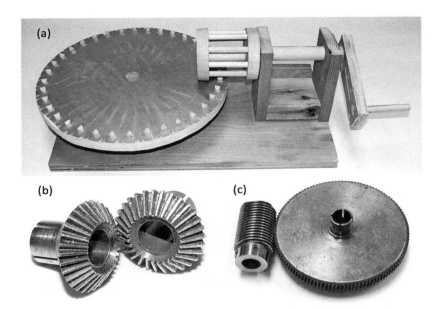

Figure 4.10. (a) A set of crown-and-lantern gears; (b) a pair of bevel gears; (c) a pair of worm-and-wheel gears, as commonly used to reduce the speed of electric motors. All three types of gears shift the axis of rotation by 90 degrees.

The modern mechanical engineer can choose among a variety of gearing systems to turn that right angle. The simplest, at least conceptually, consists of a pair of so-called bevel gears, gears (as in fig. 4.10b) with their teeth at 45-degree angles to the long axes of their shafts. If the gears have different diameters, modest changes in rotation rate can accompany the change in direction. But—and I speak from personal experience—bevel gears ask a lot in the way of alignment, bearings, and shaft rigidity as the load tries its level best to push the shafts out of position. Put another way, even when made of metal, they're distinctly unforgiving.

The designers of the fifteenth to eighteenth centuries commonly turned to what we now regard as an unfamiliar solution to the problem of turning that corner, one shown in figure 4.10a.[15] They put a crown gear on the main vertical shaft of the whim, quite a wide wheel with pegs protruding upward from its top surface, hence "crown" (even if the pegs sometimes stuck downward from the bottom). Those pegs interdigitated with the rundles of a so-called lantern gear. The combination of crown and lantern solved a whole bunch of vexatious problems:

- Alignment became much less critical. Normally the drum of the lantern simply rolls around on the top surface of the crown gear, but nothing changes very much if it moves around slightly above the crown gear.
- Turning of the gears did not simultaneously push them apart. So providing good bearings on the two shafts near the gears to take lateral forces on the shafts becomes less crucial.
- The system could be made fairly large without becoming excessively heavy. Large size reduces the force per area—the stress—exerted by a load at any point. That's a good thing when the main material is wood, even if some good hardwood bears the load. Of course, working with wood brings a suite of other issues. Fortunately, crown-and-lantern gearing deals decently with the severely directionally dependent mechanical properties of wood and at least tolerably with the humidity-dependence of its dimensions.
- Perhaps best, the use of crown gear as driver and lantern as driven ensured that the slow shaft speed of the whim would be converted to a faster rate of revolution of the radial output shaft. This last point brings up a subtle aspect of the history of technology that deserves more attention than it commonly attracts.

A horse turning a whim might walk at about two miles per hour; with a 10-foot radius, the motor formed from horse and whim will rotate at roughly three revolutions per minute, 3 RPM (or 0.05 Hz, if you prefer). Any ordinary automobile engine turns a thousand times faster, at least when you accelerate. Small electric motors turn at between 1,500 and 5,000 RPM. My lab has accumulated quite a few small motors equipped with attached gears to slow those inconveniently rapid outputs. All but the smallest reduce speed with a worm-and-wheel (fig. 4.10c), a way to reduce speeds by about tenfold in a single step. A rapidly turning prime mover whose rotation rate is reduced for its application is a common feature of contemporary technology. The motors of your blender and food processor might drive their blades directly, but meat grinders, cake beaters, ice crushers, and the like need speed-reducing gears. By contrast, half a millennium ago, prime movers turned slowly, and rotation rates commonly had to be increased, not decreased, so that they could do their intended tasks properly.

Whether one looks at steam engines with pistons, internal combustion piston engines, turbine engines using either external or internal combustion, or even electric motors, one can discern a historic progression from slowly turning motors to more rapidly turning ones. Higher pressures, better seals, better bearings, better lubrication—many factors contribute. The benefits include material and space economy and improved power-to-weight ratios. The general trend does become slightly obscured by another, the inverse relationship between the size of an engine of a given type and its speed of operation—other things being roughly equal, the smaller engine spins faster. That's not as odd as it looks at first glance. If you double the diameter of a rotor and halve its rotation rate, then the speed of movement of the outer edge of the rotor has not changed. The same goes for pistons in cylinders of different lengths—as you can easily figure out for yourself.

One instructive exception comes to mind and deserves particular examination. Old windmills, such as traditional Dutch ones, were large affairs that turned slowly. By contrast those that pumped water from the aquifers beneath the American prairies put smaller, more rapidly spinning rotors atop lightweight towers.[16] So far, so typical. But we're now making windmills with the largest rotors ever, ones that turn at rates typical of those of northern Europe many hundreds of years ago. What underlies the peculiar reversal of the normal trend is nothing more than the oddly dilute character of the energy source, the slow movement of low-density air across the surface of the earth. We could pair passive wind concentrators with smaller, faster rotors, but the alternative turns out not to be worth the trouble any more than it's worth replacing a walking horse with a galloping dog on a whim or treadmill.

Back to the world of Taccola, Agricola, Ramelli, and their contemporaries. What, exactly, did they do with all those whims, now equipped with radial driveshafts? Material abounds thanks to Johannes Gutenberg (c. 1395–1468), one of the greatest enablers of all time. And access has never been better, thanks to recent digitizing projects. So besides looking at hard copies of the works of these three, I downloaded books and other material attributed to several lesser-known figures: Jacobus de Strada (c. 1523–1588), Jacques Besson (c. 1540– c. 1576), Vittorio Zonca (1568–1603),[17] and Fausto Veranzio (1551– 1617).[18] The diversity of hoists, cranes, and pumps leaves no doubt about the ingenuity of the technology of the times.

Earlier in the chapter, grain mills put in an appearance in the form of rotary querns turned either by humans or animals. These sixteenth-century engineers drove millstones in a far more effective way. A quern turned by a poorly coupled donkey or by humans grasping handspikes could not have been at all efficient, and the time-consuming nature of grain grinding means that efficiency should be of some importance. Turning a simple quern with a well-yoked ox instead of donkey or human gains nothing beyond an easy day for the ox. The trick for making better use of this far more powerful engine consists of speeding up the rotation of the millstones, just what crown-and-lantern gear pairs accomplish. Nor were they content with the four- or fivefold increase from a single pair. As in figure 4.11, several illustrations show arrangements in which the whim, usually horse- or ox-driven, led to two sequential sets of crown-and-lantern

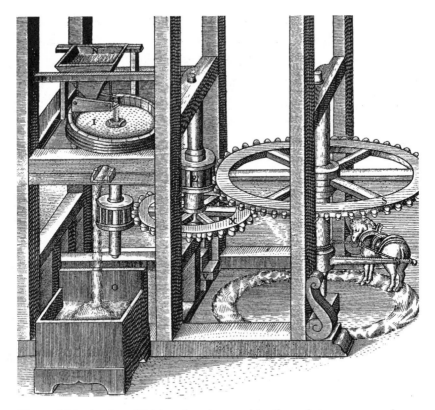

Figure 4.11. A grist mill driven by a horse-whim through two crown-and-lantern speed-increasing gear pairs (Ramelli, 1588, plate 121).

speed increasers, achieving about a twenty-fold increase in rotation rate. That would, for instance, convert 4 RPM to 80 RPM, quite a respectable rate for a large millstone. And the numbers assume no artistic license in the drawings, which I think unlikely—the arms of the whims look unrealistically short, so I wouldn't be surprised if the real crown-and-lanterns had still higher ratios. These just wouldn't have looked as tidy ("photogenic" these days) in stylized drawings—drawings produced by artistically sensitive engineers, not mere draftsmen. So the sixteenth-century (and later) miller would have had a single large, rapidly turning mill, in contrast to the miller of Pompeii, stuck with an array of small donkey querns.

Except for a human-powered hoist described by Ramelli,[19] other devices powered by crown-and-lantern gearing used only single gear pairs. Most common are pumps and hoists, which comes as no particular surprise, and they amount to a large number of variations on a few themes. The succeeding centuries saw little fundamental change in these horse- and ox-powered whims with horizontal power takeoff shafts—for the most part, they proliferated in number, geographical distribution, tasks, and sizes. The main change accompanied the gradual increase in availability, meaning affordability, of ferrous metals. In particular, that made gears and shaft work of the now-familiar kind practical, which in turn facilitated putting the radial shaft low rather than overhead. A shaft could be small enough so an animal could step over it each time it came around, or the shaft could be contained in a shallow, covered trench or conduit. Relief from power-handling overhead supports by the low-mounted radial shaft allowed lighter, portable grain mills. Of course, they needn't have been all that light since each mill had to come with at least one draft animal; portability provided an intrinsic advantage over any waterwheel or windmill![20]

The most impressive machine based on a whim geared to a horizontal shaft (to give my own opinion) happens to be a relatively early one—the hoist designed and built by Brunelleschi to lift beams of sandstone for the construction of the cathedral Santa Maria del Fiore, in Florence, Italy, in 1421. The beams, up to 1,700 pounds in weight, were much too heavy for the human-powered treadwheel cranes introduced by the classical Romans. Instead Brunelleschi used a very large whim powered by one or two oxen. Instead of the usual crown-and-lantern gear pair, he used three pegged gears (as on crown gears) arranged (as in fig. 4.12) so that the vertical shaft bearing two of the pegged gears could be raised and lowered by a screw to engage either the gear that raised the load or one that lowered the load. That way

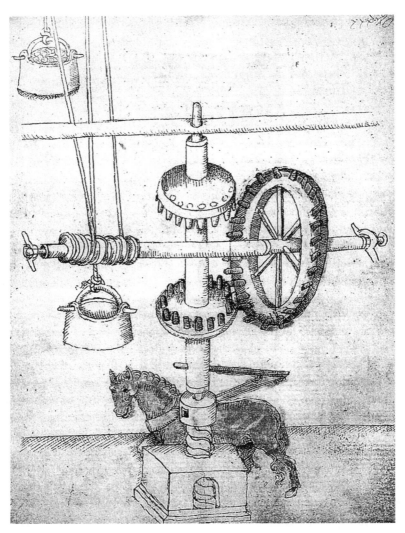

Figure 4.12. The reversible hoist used by Brunelleschi to lift beams; normally oxen drove the whim. The sketch was done by his contemporary Mariano Taccola (reproduced in King, 2000).

the oxen didn't have to be unhitched and turned around with every change in direction. The scale, again, was monumental—Pisan ship-builders supplied ropes two and a half inches in diameter that had to be kept moist to limit buildup of heat as they flexed.[21]

In addition to grinding, pumping, and hoisting, more compli-cated applications eventually drew on these whims. Around 1784 a Scottish mechanical engineer, Andrew Meikle, invented a full-blown threshing machine, which accepted both wheat and chaff, not just kernels of grain—all driven by horses turning a whim. These became ever bigger and better, reaching their consummation in the great machines of the American Midwest in the mid- to late nineteenth century. Whims also pulled plows and larger trenching equipment, where swampy bottomlands had to be drained—a whim took only a small area of firm footing and thus avoided any need for draft ani-mals to make their way as they worked through the soft substratum itself.

One further fairly widespread purpose for which we asked animals to walk in circles has drawn surprisingly little attention. If one mentions horse ferries, the usual image that comes to mind is some winching arrangement in which a horse turns a capstan and pulls a barge across a river. And, yes, that system has occasionally been used. That the horse might ride on the boat seems less likely at first glance. But boats that were at once horse-propelled and horse-carrying gave valuable service, particularly in eastern North America between about 1810 and 1860. At no stage were they anything other than ugly stepchildren to the trendy steamboats, proliferating simul-taneously, so these "teamboats" (the contemporary quip, not mine) left only a tiny historical footprint—or hoof mark. Since horses did almost every other power-demanding task, perhaps, the horse ferry lacked a sufficiently radical aspect to make it noteworthy.

Strictly speaking, they weren't even new. Some accounts claim that, around 370 CE, the Romans built a ferry in which three pairs of paddle wheels were driven by three ox whims; it is claimed to have carried troops across the Strait of Messina to Sicily. The crossing is narrow enough for that to be at least minimally possible even if the negligible remaining deck space shown in what purports to be an illustration does undermine its already limited credibility.[22] More reliable reports of horse boats and even human-powered boats, almost all one of a kind and whim-driven, span the years from about 1500 to 1800.

The general inland maritime context stretches the bounds of the present book but is worth a brief sketch. By the end of the eighteenth century, steam engines had developed from Thomas Newcomen's huge, heavy, inefficient pumping machines to lighter, more weight- and fuel-efficient devices that could conceivably move themselves— if coupled to proper propulsive equipment. Two designs of 1787 succeeded technically but failed commercially. James Rumsey very cleverly made a reciprocating steam-driven piston draw water in beneath the hull of a boat and then squirt it out in the rear. This was proper jet propulsion very much as is done by squid; unfortunately, jet propulsion turns out to be quite inefficient at low speeds, which is why squid use fins for routine swimming and why low-speed aircraft still use propellers. John Fitch coupled his steam engine to a set of paddles that emulated in appearance and action the webbed feet of waterfowl, in his case losing out due more to mechanical than fluid dynamic inefficiency.

Commercial success came in 1807 with Robert Fulton's steamer, a vessel with a pair of large paddle wheels, one on each side of the hull.[23] Fulton thus not only took good advantage of a wheel-and-axle arrangement, but he used wheels as drivers as was done in almost every non-animal-powered land vehicle that followed. The tips of his large paddle wheels barely dipped into the water, which improved propulsive efficiency—they spent a minimum of energy pushing water down or lifting it up. And he had the advantage of the deep pockets of Robert Livingston, enabling not only construction of a fairly large (and correspondingly more efficient) vessel, but the purchase of a steam engine from Boulton and Watt, of Birmingham, En-

gland, at the time the foremost manufacturer in the world. He also
ran his boat up and down the Hudson River, a route pre-adapted for
steamboat service—a long estuary-like body that led far inland from
a large city and harbor to good farmland and water-mill sites inland.
(The Erie Canal and the real hegemony of New York City came two
decades later.)

So where do horse ferries come into the picture? Steamboats,
especially the earlier ones, were not the handiest of machines. Engines were costly and required considerable maintenance, getting
up steam took time, and small boats were the least economical of
steamboats. In a way they were like jet airplanes. If you had one, you
wanted it never to sit idle. A horse ferry had exactly the opposite
characteristics. Cheap, familiar technology, best for intermittent service, and naturally small—in short, a complementary alternative.
For decades, then, steamboats ran up and down the Hudson, while
horse ferries crossed from shore to shore, connecting towns at many
points along its length. The same combination served on several
other rivers of eastern North America as well as long, narrow lakes
such as Lake Champlain.

Horse ferries have finally acquired their Homer—better, Homer
plus Schliemann. Between 1990 and 1992, a team led by Kevin Crisman and Arthur Cohn carefully excavated the remains of a horse
ferry that had been scuttled in Lake Champlain around 1860; they
wrote a book describing the excavation, the boat, and the whole era
of horse ferries and their operation: *When Horses Walked on Water*.[24]
Of the three designs that saw extensive use, the first one has the most
relevance to the present chapter—we'll talk about the other ones in
the next chapter.

This most basic horse ferry consisted of the kind of whim we've
been looking at, now mounted on the deck of a boat. Typically, the
boat had a pair of hulls, with a paddle wheel between the hulls and
beneath its deck, and the whim often placed in a small house to keep
the horse out of the weather, as in figures 4.13 and 4.14. The shaft of
the whim descended down through the deck and drove the paddle

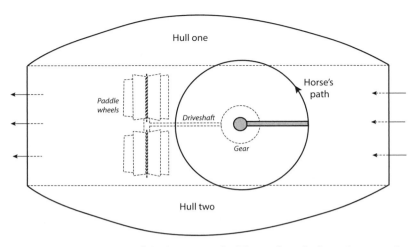

Figure 4.13. A drawing of the basic twin-hull horse ferry looking down on the deck—with an admittedly large amount of my own guesswork since I could find no detailed plans or exact description. Dotted lines and italicized words refer to items below the deck.

Figure 4.14. An old picture of a paired-hull horse ferry, this one used to cross Halifax harbor in Nova Scotia (from Dartmouth Heritage Museum, Dartmouth, NS, as reproduced in Crisman and Cohn, 1998).

wheel through a set of bevel gears. The particular arrangement of gears seems to have varied from boat to boat; most were built by local foundries and boat builders who left no records. Inasmuch as both undershot mill wheels and horse whims were long-established technologies, these ferries needed no notably novel mechanical elements.

The particular design, though, had at least three disadvantages. One, noted by Crisman and Cohn, was that the horse whim took up too much deck space. Naturally deck space represented payload—people, herds of animals, wagons, and so forth. Another is that a horse represents a considerable load, one that awkwardly moves back and forth from port to starboard, rocking the craft. The third disadvantage is more subtle and may not even have been widely recognized at the time. A paddle wheel works best if it's large and dips only its tips into the water—as did those of the Fulton steamboat and its successors right up to the lovely ones on the Hudson in the 1940s, ones I rode on as a youngster with my grandfather. Otherwise, as already mentioned, too much energy is lost trying to push water down as a paddle blade enters the water and lifting water up as it leaves the water. An under-deck wheel needs to be a small, wide wheel, not a narrow, broad wheel, and thus it's condemned to be either a very slow pusher (dipping too little area in the water) or inefficient (dipping too much). In general, paddle wheels never achieve the efficiencies of propellers, even the efficiencies of the hydrodynamically poor propellers of the late nineteenth century—side-wheel river steamers persisted mainly because their ability to reverse one or the other wheel made them quite wonderfully maneuverable.

[5]

Or Being Encircled

Another set of devices complements the whims of the previous chapter. Just as Copernicus showed that one could take advantage of alternative frames of reference—in his case a heliocentric rather than earth-centered solar system—humans have often chosen to make the mill go around the animal instead of asking the animal to go around the mill. The items involved are all those treadwheels and treadmills, of diverse, ingenious, even fiendish contrivance. Contrary to assertions on the websites of several purveyors of contemporary cardiovascular treadmills, both treadwheels and treadmills greatly antedate those of a sadistic early nineteenth-century British penologist.

First, we need a look at the basic arrangements, of which there seem to be four. The sequence below may be easiest to describe, but it pretends no claim of chronological validity.

- Closest to the whims of the last chapter, the animal can be harnessed so it pulls or pushes on a radial beam, but now the radial beam is off center and fixed to an external support, as in figure 5.1a. What moves is a platform beneath the animal's feet rather than the animal itself. If the animal is large, a horizontal platform usually works well enough, but for smaller animals, tilting the platform permits extraction of more work. From what we (now) know of the way the cost of ascent depends on the size of the animal, the smaller the animal, the greater will be the optimum tilt.

- Most of us are familiar with the little wheels in which small pet rodents can take exercise—often called "hamster wheels" (fig. 5.1b). In one of these, an animal runs within a hollow cylinder turning on a horizontal axle as it makes a futile attempt to run forward and climb upward. You can now buy larger versions of such wheels to exercise your family dog or even to attempt to exercise your cat. If you have a small rodent and such a wheel, you might try connecting the axle to a tiny electric motor (here working as a generator instead); connecting the output to a flashlight bulb makes the animal do some observable work.

- With only slight modifications, the wheel can be designed so the animal's futile activity consists of climbing up the outside rather than the inside of the wheel or else walking on its top rather than its bottom (fig. 5.1c). That allows a smaller wheel, but as a practical matter it restricts its use to bipedal—and cooperative—humans. It has another problem that will become apparent further along.

- Technically the most difficult and therefore the most recent incarnations are the treadmills with which we're most familiar, the ones (as in fig. 5.1d) in which a rubber mat slides across a tiltable metal plate or other support, usually driven by an electric motor. We have big, soft feet, and we don't worry about the friction of the mat on the plate since we're not trying to produce a proper power source. Your efforts on the treadmill may very slightly lower the electricity its motor asks of the electric company, but who cares?

I can't decide whether I'd rather work out on a treadmill (or treadwheel) or a whim. Both are consummately boring activities, my

Figure **5.1.** Treadwheels and treadmills. (a) Fixed animal position on the radius of revolvable turntable; (b) a treadwheel with the animal climbing on the inside; (c) a treadwheel with the animal climbing on the outside; (d) a moving belt treadmill—an obviously more recent drawing. (a) and (c) are from Agricola; (b) comes from Ramelli; (d) graced an issue of *Scientific American*, in 1880; the horse drives an electrical generating unit.

opinion of the treadmill a matter of much personal experience. The whim would give a slightly more varied outlook but at the same time would make you a little dizzy, so it would be tolerable only if the arm on which you pulled or pushed was quite long. If nothing else, that requirement for acreage would limit its appeal for contemporary exercise facilities. Conversely, a whim has to be the easiest of these devices to build as a do-it-yourself backyard or basement contraption.[1]

While the ancient cultures of the Middle East don't seem to have used treadwheels or treadmills, the classical Greeks and Romans certainly did. Two tasks in particular received their power—hoisting heavy stones in construction projects and pumping water, this last both for irrigation and for draining mines. We know more about the specific machines of the Romans than of the Greeks, mainly because a large amount of the writing of one excellent Roman source, Vitruvius,[2] survives. But the two cultures were technologically distinct in no major way. Again, the historian of technology suffers under the combined burdens of devices produced largely by illiterate craftsmen, of the loss of what limited written material may have been generated, of unfamiliar descriptive terminology in what writings we do have, and of an absence of diagrams and other visual aids. We also have some uncertainty of the precise sizes of ancient measures, but this may be the least of the problems, affecting only quantification.

Humans provided most of the prime movers for both lifting solids and pumping liquids.[3] Lifting, perhaps of less importance overall, relied on one particular device, a treadwheel in which one or more persons climbed, hamster style. A rope around an extension of the axle led upward to a pulley or some more complex arrangement of pulleys and then down again to a basket, sling, or grapple for the load. The arrangement had numerous virtues. It wasn't excessively massive. Human engines could respond quickly to commands. The diameter of the axle extension, the windlass, could be chosen to match human force and power outputs to the desired speeds and loads. More subtly, climbing inside a wheel produces a nicely self-regulating system. When only a light load taxes the system, the person can walk nearly on the level, but as the load increases, the person will find it necessary—and almost automatically begin—to climb ever more steeply to maintain the same speed. Ample evidence of such tread-wheel cranes remains both in written form and as bas-relief on several monuments, from one of which figure 5.2 has been taken.

Pumps of any kind trade off two principal variables as they apply their power to their tasks—how much fluid they pump and how much pressure they impart to it. For water or air pumps, that means volume per time (gallons per minute or some equivalent unit) and the height the water is lifted, respectively. The product of these two variables is the power that the pump puts out. Thus for a given maximum power input, more flow means less pressure, and vice versa. One basic class of pump does well for low-volume, high-pressure applications, appropriate for tasks such as inflating the tires of a racing bicycle. Another class does well for high-volume, low-pressure applications, as might work well for a pump intended to inflate an air bed. If you're trying to drain a deep mine, height obviously matters a lot, and you need a pump capable of generating sufficient pressure to push water to a height equal to the depth of the mine. Otherwise it will be quite useless, at least unless several are arranged serially. Conversely, if you're trying to move water from one rice paddy to an adjacent one, then you worry about volume moved per unit time, and you'd find a pump specialized for lifting to great heights quite inefficient. In power-limited technologies, efficiency counts for a lot; a human hard at work with good equipment, remember, will deliver only a modest 100 watts.

The best high-volume, low-lift pump of classical antiquity was the Archimedean screw pump, basically a big helical screw in a casing much like the pusher in the middle of an old-fashioned meat grinder (fig. 5.3a). In the ancient version, the screw and its casing turned together. This turning moved water up an incline in a way similar to the way the helically grooved shank of a fluted drill bit moves wood chips up and away from the cutters at its tip. Whether Archimedes

Figure 5.2. Marble relief, from the tomb of the Haterii, later first or early second century CE. The Haterii were a famous family of building contractors in Rome, so the image is appropriate. The large treadwheel at the bottom left appears to be driven by at least half a dozen men climbing the inside of the wheel.

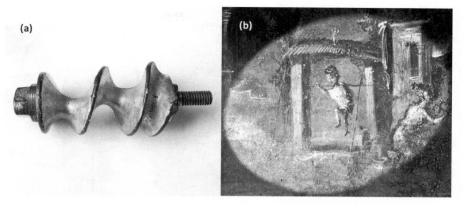

Figure 5.3. (a) The driving helix from our well-used household hand grinder. We later bought an electric one, but the screw (made of plastic) was too shallowly cut, so it worked poorly. (b) A fresco from Pompeii (http://kattler.dk /schiolers/uk/152.html). Look carefully and you can see water pouring into a jug, losing in the waterfall most of the height it gained moving through the pump in this artistic rendition.

invented the pump remains far from certain, indeed unlikely; but it works well and remains of value, particularly for large projects where clogging with debris might disable other kinds of pumps. It also gives good service for friable solids—gravel, seeds, and so forth; the ice maker in our refrigerator moves the ice "cubes" forward toward the dispensing funnel with one. Modern water-pumping versions designed for a single operator depend on a now-ordinary crank at the upper end of the screw. But, as noted, cranks were uncommon in antiquity, and cranked Archimedean screw pumps go back only to the fifteenth century. So how could a person power such a pump? We have several surviving figures, a fresco from Pompeii (fig. 5.3b) and a terra-cotta now in the British Museum.[4] The whole pump forms a long, narrow treadwheel. A horizontal bar a few feet above the pump provided what must have been a crucial support for an otherwise awkwardly precarious perch. One suspects a system of no great efficiency, but these pumps were apparently in fairly wide use nonetheless.

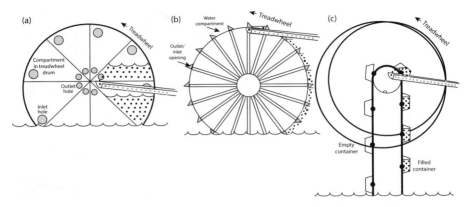

Figure 5.4. Three Roman treadwheel-driven water-lifting devices. (a) A drum, lifting to a little less than its radius; (b) a high-lift drum; and (c) a bucket-chain, good for still greater lifts, if with still less volume flow. (From Vitruvius's sketches in Book X, redrawn with guidance from text in Landels, 2000.)

One trade-off that gave higher lift (with relatively less volume) for a given power input into the task was a tympanum, as described by Vitruvius—a big, flat, vertical drum,[5] as in figure 5.4a. Well known, it seems, to both Greeks and Romans, the version he described divided the drum around its circumference into eight compartments. Each of these had a slot where it first entered the water, so the compartment acted as a scoop. It also had a hole near the wheel's center, so when the compartment came up to a nearly horizontal position, water would drain out. In that way, with each turn the drum would sequentially lift the contents of each of the compartments to almost its own radius. And thus lifted, the water would then run out through a channel leading to whatever it was intended to supply—even to a trough beneath another such pump mounted higher up.

Power for such drums came, as with Archimedean screw pumps, from their direct use as treadwheels—usually one or more people climbed monotonously on the descending side of each tympanum. Alternatively, a second wheel on the same axle may sometimes have served as treadwheel, perhaps with the operators inside instead of outside. Walking up the outside of a wheel, although mechanically simpler, suffers from a slight disadvantage over walking up the inside, an issue alluded to earlier. For an inside walker, an increase in load slows down the wheel, which almost automatically leads the walker to walk forward and thus to climb a steeper part of the wheel, thus putting out more power (or going slower) and offsetting the increased load. So the system has a nicely self-regulating character, at least up to the power limit of the walker. Walking up the outside does not. If an increase in load slows the wheel, the walker tends to climb onto a less steep portion and is all too likely to go (literally) over the top. So power output diminishes rather than increases, quite the wrong response to an increase in load. For the steady loadings of water-lifting, and for relatively experienced wheel-walkers, this might not prove a serious disadvantage, but it does make the basic device less versatile.

For still higher lift, a similar drum came into play (fig. 5.4b), one that lifted water almost to its full diameter but raised less water with each revolution. It deserves a few words of admiration. It may have been little more than a series of small compartments, but the design of the compartments showed considerable ingenuity in the way it at once maximized delivery of water, minimized spillage, and provided cleats for the feet of the treaders. In one of the complex Roman mines at Rio Tinto in Spain, eight pairs of such treadwheels in vertical series raised water no less than 97 feet.[6] A variant of this device had its buckets attached around the periphery of the wheel, oriented so each scooped up water at the bottom and tipped the water into a trough when the bucket passed over the high point of its orbit. That precluded simultaneous use as a treadwheel, so it required either a separate treadwheel or an animal-powered whim with a horizontal power takeoff shaft and crown-and-lantern gearing. It was apparently less common than the simple scooping drum.

A bucket chain gave the highest lift in a single stage; at least in theory it could reach any height (fig. 5.4c). In the version Vitruvius describes, the axle of a treadwheel extended outward on one side to engage some kind of sprocket or high-friction region around which the bucket chain could be turned without slipping. With good design of buckets, spillage could be reduced to insignificance. The main difficulty with the pump, noted by J. G. Landels, was cost—it required more metal parts than the other treadwheel pumps if it were to run for long periods. His other complaint, the low output, boils down to nothing more than a trade-off unavoidable in any pumping system. For a given investment of power, a high-lift pump will of necessity have a low volume output.

What changes would we see if we looked again at treadwheels a thousand or so years later, say in the sixteenth century? We can again take the rich collection of mechanical compendia described in the last chapter as illustrative of the technology of the period. For simple lifting devices—cranes—not a lot had changed beyond increases in the sophistication of gearing and general construction. The basic treadwheel with one or more humans inside had been used for centuries in the construction of the great Gothic cathedrals of Europe; it either was reborn, perhaps from surviving bas-reliefs, or maybe it had never truly been lost. Brunelleschi's particularly elaborate (and especially powerful) riff on the machine, mentioned earlier in connection with building Il Duomo, appears to have been one of a kind.

Nonetheless, substantial change becomes evident if we take a wider look at the ways in which people and animals acted as prime movers while themselves staying in place. A careful look can even recognize that some items no longer appear. Amid the remarkable diversity of arrangements included in these hundreds of fifteenth- and sixteenth-century figures, one group of devices in particular seems no longer to have any significant role. We have no problem declaring its loss no loss at all—"progress," to apply that most loaded of terms. These missing designs are all the versions of pumps in which the prime mover, usually a human, works on the outside of the wheel—the treadwheel or drum-based pumps and the classic version of the

Archimedean screw pump. In my opinion, the most likely culprit is that problem of operational instability mentioned previously.

The nearest these designers approached externally driven treadwheels consisted of two schemes. One put humans on seats so they had to work in a fixed position, in practice either with their feet pushing a turntable or else repeatedly grasping handspikes. The other drove treadwheels with quadrupeds but enlisted only two of their four legs. Agricola, for instance, shows a horse forced to stand so close to a treadwheel that it has to press down on the slats of the mill with its forefeet[7] (fig. 5.5, *left*); the particular arrangement drove a set of bellows that ventilated a mine. Strada has his horse's hind legs protruding through a shallow gap in a floor onto the top of a treadwheel; a wall (and hay dispenser) prevented it from going farther forward[8] (fig. 5.5, *right*). In both cases the animal would have been sufficiently constrained to offset any instability.

Curiously, in my perusal of quite a few fifteenth- and sixteenth-

Figure 5.5. Quadrupeds on bipedal treadwheels. On the left is Agricola's bellows-pumping horse who drives a treadwheel with its forelegs; on the right is Strada's grain-grinding horse who uses his hind legs to drive the treadwheel. Both animals are eating, so to speak, on the run.

century compendia, I found not a single Archimedean screw pump. After all, we still make use of it, and we do so in very large-scale installations. Perhaps a pump of a diameter that can be constructed without too much difficulty would not be wide enough to make good use of a human-size animal operating it as a direct treadwheel, resulting in poor coupling for transmission of power. Quadrupeds might be used, but only if either front or hind legs were supported on some non-moving platform, which sounds awkward, but of which we have just noted examples. Of course, one can't dismiss the possibility that its performance as a high-volume, low-lift pump was not what contemporary tasks required—for instance, mine drainage rather than maintaining slightly different water levels of adjacent rice paddies.

Nor did I find any fifteenth- or sixteenth-century examples of either form of the vertical drum pumps used by the Greeks and Romans. Even if not driven directly by people treading on their outsides, their axles could have easily been extended to those of inside-operator treadwheels. So that particular instability can't explain their absence. A more likely explanation is the much greater use of bucket-and-chain pumping for high-lift applications, and that's probably a result of the greater availability of metals. Agricola, the mining engineer, shows nothing like the multiple stages of drum pumps used by the Romans in the Rio Tinto mines in Spain—his chains can go as deep as he wants in a single stage. Bucket-and-chain only touches on the sophistication achieved. Along with chains come pairs of buckets, one ascending while the other descends, moving compartments, in which disks or ellipsoids travel downward and upward again through close-fitting pipes, and a host of multi-cylinder piston pumps.

The story would be particularly tidy if one could assert that externally driven treadwheels had passed from the scene, never to reappear. But that's not quite the case. One author, Fausto Veranzio, still used them, now for turning millstones, pressing oil, and dredging—but with platforms adjacent to the steepest part and, in one case, with a top grab bar to further limit climbing onto a less

Figure 5.6. Extracting power with turntables. At left is one of Agricola's two-person, level turntables, here working as a hoist; at right is Ramelli's tilted turntable (1588, plate 123), operating a grinding mill.

steep portion.[9] As we'll see, they took a final bow in the nineteenth century in a wholly different context.

One fully novel type of treadwheel appears at about this time. I find no mention of it in the earliest work I looked at, *De Ingeneis*, by Taccola, contemporary and good friend of Brunelleschi. But it appears in quite a number of drawings by Agricola, Ramelli, and Strada, always powered by one or two humans, as in figure 5.6 (*left*). This is the horizontal, rotating plate with a central, vertical axle—like an old-fashioned record-playing turntable. Completing the setup is a radial bar (or equivalent) that provides a grip for the person, who turns the plate by trying to push the bar forward. So simple, and so much more compact than a high vertical treadwheel or a wide whim—if perhaps a little less powerful. But . . .

A turntable works well if carefully balanced, for just the same reason that we take pains to balance the wheels of our cars. Put a person on one side of a turning wheel, and it will be anything but well balanced. The disk and the central shaft will want to bend under the person's asymmetrical loading, so both need to be exceedingly rigid. As bad or worse, the rotating surfaces around the top of the shaft that rub against its housing will be subjected to severe lateral force—precisely in the hardest place to add lubricant and in the orientation in which the lubricant is most likely to leak out. Once again, all turns on bearings. Which means metals, which the Greeks and Romans were loath to purchase, at least for civilian purposes. (Yes, as mentioned earlier, lignum vitae wood might have worked wonders, but the tree grew only in the New World. Upon discovery by Europeans, in the sixteenth century, the wood was mainly seen as a [futile] remedy for that purportedly New World disease, syphilis.[10]) A partial amelioration of the problem of lateral shaft loading would be to add rollers out near the periphery, but these add yet other complications such as ensuring two accurately parallel surfaces, more bearings and axles, and so forth. Of course, for balance two drivers on opposite sides are far better than one. But they still walk with periodic footfalls, varying the load and imparting vibration.

Ramelli shows an interesting variant of this horizontal turntable, one inclined by about 25 degrees (fig. 5.6, *right*).[11] The incline became increasingly common with time; by the nineteenth century, it seems to have become the standard arrangement, both for human-driven and for small animal turntables. That precluded paired drivers, since one would now walk downhill. But it makes good sense from a biomechanical viewpoint, both in terms of long-term skeleto-muscular comfort and of power efficiency. On a tilted turntable, one in effect goes steadily uphill, just as in a treadwheel, applying one's own weight to advantage and doing so in a posture common to daily activities—one for which we're decently adapted. Whether Ramelli's version came close to the optimum angle for a human remains uncertain. For mountain paths, 14 degrees, not 25 degrees, marks the optimum for gaining altitude with least useless horizontal travel,[12] but not only is the applicability of that datum to the present issue questionable, but for illustrative purpose Ramelli's drawing almost certainly exaggerates the actual angle.

In any case, two devices, the horizontal treadwheel with the animal inside and the horse whim, were the default stationary prime movers where wind- and waterpower were not available. What animals powered these fifteenth- and sixteenth-century engines? Humans, horses, and oxen did most of the work. Agricola has several goats working inside a vertical treadwheel that powered a gristmill. Allusions to bears appear here and there, and some populations of European brown bears are smaller and arguably more tractable than any of the North American bears. Bears did figure in the culture of the time — special arenas in Elizabethan England featured bear baiting, a sport (so to speak) in which bears were set upon by dogs. Donkeys, sheep, and dogs may also have occasionally been pressed into service on treadwheels.

Nothing particularly revolutionary appears to have happened to treadmill technology during the three hundred or so years that followed. The technology as a whole may just have receded as anything worthy of great literary or illustrative attention. Whims, the central players in the last chapter, continued to perform their crucial if unexciting role, while the big vertical wheels became relatively less common. But, as in so many other aspects of human technology, the nineteenth century proved to be a time of major change. Perhaps the changes did not reach the same transformative character as the introduction of self-transporting combustion engines and instantaneous communications, but their scale was at least reminiscent of that around 1500.

I alluded earlier to a new context for externally driven treadwheels. If one looks at an English-language dictionary or book of quotations from the late nineteenth century, treadmill carries a clearly penal connotation. Badly behaved inmates were asked (putting the request euphemistically) to serve time on a treadmill of this kind, at least in English-speaking countries such as Britain, its colonies, and, less commonly, the United States. Even now, the great *Oxford English Dictionary*[13] recognizes no other meaning (even if we include metaphorical allusions) for treadmill. All the sources with which I'm familiar emphatically declare that prisoners found the labor extremely unpleasant and note its effectiveness in inducing positive behavioral adjustment. The typical schedule consisted of a two-thirds duty cycle, with an overall time-averaged power output (depending, among other things, on the weight of the climbers) of 70 to 90 watts—hard labor by both old and new measurements. The wheels themselves, as in those of figure 5.7, were very long affairs, designed to accept a large number of individuals at once; their power pumped water, ground grain, and did other useful tasks.[14]

What about that intrinsic instability of externally driven penal treadmills? Almost all illustrations show a type of grab bar near the top, which will automatically position the inmates. A little improvement in one's situation might still be possible by holding the bar opposite a lower part of one's body, thus treading nearer the top. Again, anyone who puts on a slight burst of speed and goes forward toward the top—or even a little over the top if that's possible—is

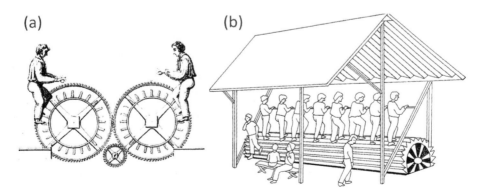

Figure 5.7. Two penal treadmills. (a) The original design of William Cubitt, showing only two of a series of treaders (from Cubitt, 1822). (b) One in use at Bellevue Prison in New York City (redrawn from Hardie, 1824).

rewarded thereafter with an easier time. The laggard has to work harder, which will tend to make him (or her) lag further. One wonders about the consequent rules, interpersonal interactions, and levels of supervision.

Not only did the nineteenth century see substantial change in treadmill technology, but it marked an end "with a bang, not a whimper," to transpose T. S. Eliot's phrase. Nowhere was that more striking than in North America. Why there? I don't prefer the intrinsic explanation—what has long been called "Yankee ingenuity"—but rather think extrinsic economic and geographic factors are more likely responsible. What differed in nineteenth-century North America were population levels and patterns of land use, of great importance back when most people either farmed or lived by occupations only a step removed from farming. Land was more widely distributed among the populace; at the same time, land holdings by families who actually worked the land were larger—family farms predominated rather than estates with hired, indentured, or peasant labor. And these farms were large relative to the numbers of working individuals even in the large families of farming communities. Robert Frost has a poem about the death of "the hired hand"—note, one hired hand! The situation thus provided unusual impetus for the evolution of all manner of labor-saving devices, both self-powered and those requiring external power.

Eastern North America also had relatively widely distributed sources of waterpower, in the form of creeks that ran down into often-navigable rivers, and both coal and iron ore; small deposits of this last long ago ceased to be of commercial concern. So fabrication and distribution of the relevant farm machinery presented few problems. North America also went further than elsewhere in promoting literacy, and its people were more mobile than those of most agrarian societies—many had only recently settled wherever they were. In short, here lived the least "traditional" of traditional agrarian societies.

All of which "pre-adapted" North America for accepting a re-markable diversity of animal-powered contraptions, whether horse-drawn fully metallic plows, reapers, balers, rakes, and so forth, or horse-whims (as often as not called "sweeps"), or two other kinds of machines in which the animal held position while driving some rotational element.

The first of these we've already met, the tilted turntable as shown in one of the illustrations of Ramelli. You name it—small-scale, re-petitive household and barnyard tasks were driven with great ingenu-ity by tilted turntables powered by, again, you name it—sheep, goats, dogs, and children. Drawing water, churning butter, and lots more mundane tasks were thus driven. Inexpensive ferrous metals mini-mized the problems of bearings, friction, balance, and wear. Most often turntables were topped with wooden platforms, which drew on the most familiar of materials to produce something that gave first-rate service. No part required high precision in manufacture or unusual materials suitable for high temperatures or that needed special lubrication. Solutions to the remaining problems of volition and the size of the unit engine, though, awaited combustion engines.

The second type represented real novelty. This is the treadmill as we know it from fitness centers, cardiac rehabilitation clinics, and sporting goods stores. It draws a continuous (circular) belt across some flat platform, with that portion of the belt returning by going in the opposite direction beneath the platform after passing over rear and finally front rollers. While rubber does the job on all present-day belts, one or another form of interdigitating metallic chain did the job on early ones—sometimes with the addition of replaceable transverse wooden slats. The chain-work had to run smoothly across an array of rollers with a minimum of friction as well as with a min-imum of give, doing both while struck by the feet of the driving an-imals. Referring to "early" ones brings up the question of date of origin. About as precise as one can be is to point to the 1820s for early belt treadmills.[15] Rubber as we know it traces to Charles Goodyear's patent for vulcanization in 1844; the combination of rubber and internal fibers (as in automobile tires and treadmill belts) came much later.

Still, one form of wooden-slat treadmill survived up to at least the 1960s, giving continuous service in hard use. Escalators then took the form of a series of something close to steps, each one formed of a set of lengthwise wooden slats, with about the same slope for all the sets on which riders stood. The spacing between slats permitted them to interdigitate as they ran over the main top and bottom wheels. One could not slip down from one step to the next as on modern escalators, so they were probably quite safe even when packed with rush-hour commuters.[16]

Horses and oxen powered most of the treadmills of this kind, which thus had to be capable of withstanding repeated impacts from their hooves, the least forgiving of all mammalian feet. At the same time, the treadmills had to be designed not to injure the feet of either kind of animal. Whether the treadmill should be inclined generated considerable debate; if the platform was not inclined, then the animal needed harnessing or yoking—although that was a familiar enough practice. Portable treadmills came with the territory since the power source could just as easily pull the whole thing as propel its belt. So enormous numbers of treadmills were built and performed a great variety of functions on nineteenth-century farms. They came in all sizes, from small ones that could be powered by dogs, sheep, or even children, as in figure 5.8, to the bovine or equine versions, as in figure 5.9. Many, perhaps most, acted as general-purpose portable power sources similar to early farm steam engines and tractors—leather belts led from treadmills to whatever needed their output, with relative pulley sizes providing a proper match of speed and power. On some treadmills, several animals walked side by side, but most treadmills appear to have been single-animal units.

But we shouldn't regard treadmills, turntables, and whims as purely or even mainly agricultural appurtenances. Urban construction

Figure 5.8. A dog-powered butter churn, in which the moving belt turned a flywheel with a pitman arm (about which more later) that rocked a butter churn.

Figure 5.9. A typical advertisement for an inclined treadmill, this one a two-horsepower grain-grinder, from the mid-1800s.

sites depended on horses not just to pull wagons directly, but to pull them up slopes via cables over pulleys, in no way different from what Ramelli illustrated hundreds of years earlier. Horses powered brick-making machines, both tempering the clay and pressing the bricks. They ran hoists, sawmills, pile drivers, dredgers, cotton mills, machine shop lathes, grain elevators, pumps, and, of course, factories that made treadmills and turntables.[17]

Which brings us back to that peculiar technology of the half century between about 1810 and 1860: horse ferries. In the last chapter, horse-powered whims made their appearance, with due appreciation of the deck space on a boat that a whim co-opted. For a horse ferry, any alternative in which the animals remained stationary therefore offered an especial advantage. Both treadwheels and treadmills gave just that, and both gave good service for what we now might call second- and third-generation horse ferries. Whims, then, provided proof of concept, at least in economic terms; entrepreneurs could now look toward more sophisticated designs—"Yankee ingenuity" once again, if you like the slogan.

Conceptually, the simplest fix consisted of adopting a covered turntable rather than a whim, as in figure 5.10. The turntable may have been wide, but all of it except a pair of cutouts, one on each side, could be located just below the broad deck of the ferry. Not that the solution didn't raise some other difficulties—a deck strong enough to support animals and loaded wagons but broad enough to span the turntable had to be fairly substantial, meaning fairly well braced or thick. That meant the horses would have to work within not merely slots but within wells dropping significantly below deck level. Furthermore, a turntable wide enough for a pair of half-ton horses to develop proper power while walking had to be very stiff and well provided with bearings. Some of these turntables ameliorated these last problems with a set of small wheels that rolled around on a track out near their peripheries. These, then, could bear the extra

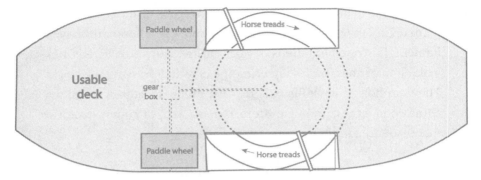

Figure 5.10. A horse ferry from Lake Champlain, showing the platform over the turntable and the gearing system that drove the twin paddle wheels. (Basic dimensions come from Crisman and Cohn, 1998.)

load imposed by any imbalance between the horses. Alternatively, radial bracing often stiffened the track. At least the boat didn't roll as a horse moved from one side to the other.

The horse ferry whose remains have been salvaged from Lake Champlain is of this last type—a turntable with radials beneath it rather than with auxiliary peripheral wheels. On it, the horses walked on a circle a little less than 20 feet in diameter. On the underside of the turntable is a circle of gear teeth about 12 feet across. Turning the turntable thus turned this gear, which then turned a lengthwise driveshaft through a conventional (and relatively small) spur gear, as in figure 5.10. That shaft leads to a set of bevel gears—which one was engaged determined the direction the side-wheels would turn. One can't speak of fore and aft or stem and stern since the boat could go either way, and the shift system permitted reversing without needing to redirect the horses. The side-wheels, incidentally, were about 7.5 feet in diameter, larger than would have fitted easily beneath the earlier double-hull whim-driven boats—and were therefore probably more efficient. They would have turned at a much more rapid rate than did the turntable.

Of the final type of horse ferry, we have fewer archaeological artifacts but considerably more illustrative material. Once one has available the pair of treadmills of the kind that were coming into common use, the rest of the design follows in straightforward fashion, so little need be said aside from reference to figure 5.11. As with the turntable boats, some kind of shelter was typically provided for each animal. Most likely the two side-wheels turned on the same axle in order to avoid any imbalance from differences in ability or motivation of the two engines. But a split in the axle together with a proper set of gears would have permitted the wheels alone to steer the boat, turning it (as did steam-powered side-wheelers) within its own length.

Horse ferries may receive next to no historical notice, but if common casual references tell us anything, in their day they were nothing if not ordinary. As put by Ulysses Grant, near the end of chapter 1 of his *Memoirs*, "I kept the horse until he was four years old, when he went blind. When I went to Maysville [Kentucky] to school, in 1836,

Figure 5.11. A twin-treadmill horse ferry, this one from Toronto, Ontario. The horses, unusually, were not protected from the weather; either the boat ran seasonally or some housing was used in the colder months. Incidentally, with horses in paired stalls, one can distinguish treadmill boats from turntable boats by noticing whether the horses faced in the same or in opposite directions.

at the age of fourteen, I recognized my colt as one of the blind horses working on the tread-wheel of the ferry-boat."[18]

Yet again, a dog that didn't bark. One kind of muscle-driven boat with a rotational engine gave good service for centuries on the rivers of China. These were paddle-wheel boats driven directly by human-powered treadwheels within the hulls—the number of wheels varied from, by reports, two to twenty-two pairs.[19] Fifteenth- and sixteenth-century Europe had undershot waterwheels in profusion. Putting power in instead of taking power out makes these into paddle wheels. The mechanical compendia already mentioned give several examples of grain mills powered by waterwheels mounted between boats tethered in moving streams. One can find illustrations of improbable-looking paddle-wheel craft with central cranks for input power, attributed with equal improbability to the late Roman Empire.[20] Yet even the more imaginative and less immediately practical of the authors, Leonardo da Vinci and Veranzio, show no treadwheel-powered paddle-wheel boats. The great sinologist Joseph Needham and the great student of medieval technology Lynn White emphasize how much the West learned from Chinese culture and technology and how early that began, so one is left wondering why, with all the pieces on the board, treadwheel-powered paddle boats were never assembled in the West.

A speculative note on paddle-wheel boats, if one of uncertain relevance here. I commented earlier on the greater efficiency of large paddle wheels that dipped only a small fraction of their radii into the water. There's another related problem in making paddle wheels drive boats, and it goes far to explain the much longer and more extensive use of paddle-wheel propulsion on lakes and rivers than in coastal and transoceanic shipping. Propellers are more efficient than paddle wheels, even the hydrodynamically crude flat-plate propellers introduced by John Ericsson in the mid-nineteenth century. Even better, propellers are far, far more tolerant of waves and of the consequent rolling, sagging, and hogging of ships' hulls; except under the most extreme conditions, propellers can be counted on to stay completely submerged and thus properly functional. That may have been of particular importance for hybrid ships, ones that also carried sails to conserve fuel. Paddle-wheel steamers did cross the Atlantic, the first non-sailers to do so, but their era of dominance was a short one,[21] whereas they enjoyed a long history as riverboats.

Perhaps geography created the key problem during the period when treadwheel-powered paddle-wheel propulsion might have appeared in Europe. The Mediterranean is a big and ill-tempered sea fed by few major rivers; if it provided the impetus for building boats, perhaps paddle-wheel boats, however powered, simply offered little attraction. One does wonder about the Rhine, Danube, and so forth. Here we bump into a basic problem that afflicts all historians but, at least ideally, no experimental scientists such as this writer. History happened only once; it cannot be repeated with slightly different input variables.

[6]

Grabbing Again and Again

Consider some of your most ordinary, everyday activities . . .

- Twisting off the cap on a bottle
- Screwing in a lightbulb
- Turning a can opener, a pepper mill, or a corkscrew
- Tightening a screw with a screwdriver, or a nut with a wrench
- Turning the tuning knob of an old-style radio
- Resetting the time on an analog clock
- Turning on the outdoor water faucet and the old hose nozzle
- Adjusting the diagonal rod that keeps the screen door from sagging
- Attaching a camera to a tripod
- Focusing your binoculars
- Turning the steering wheel of a car

For some readers, the list might be considerably longer—assuming a correct guess as to their common element.

But never mind guessing—the chapter's title suggests what they have in common. Each of the tasks requires that you rotate something farther around than you can turn fingers, hands, or arms. And for each, you adopt the same solution. You grasp it, turn it, release it, rotate your fingers (and perhaps your hand and arm) in the opposite direction, grasp it again, turn it again, and release it again, until you've turned it far enough to accomplish whatever the task demands. Grasp and release, grasp and release . . . Doing this, you're aided by enough friction between shaft and housing of what you turn so it neither freewheels nor snaps back while you reposition your grip.

Surely no trick for making one of us muscular engines achieve continuous rotation can antedate this one. Roll a stone, a log, a trussed carcass without benefit of ropes or other tools, and you'll push, release, push again, release, and so on. So it's old; it's also about the simplest and most common of all the mechanisms we'll encounter. But what can be said about its history? With some temerity (and hubris), I suggest that two trends can be discerned, one operative on a time scale of a few decades, the other extending over many millennia.

The first trend, recent and clearly ongoing, reveals itself in the allusions to "old," to "old-style," and to "analog." Looking around at the various appliances in my house, I find that most things requiring or even permitting more than a full turn have been with us for ages. Recent acquisitions have buttons or touch pads or bendable plastic flip tops or spring-loaded hinged caps, or bayonet mounts, or slide-in sockets. Why this shift? For one thing, twisting takes time, caps get mislaid, and materials tolerant of repeated flexing provide an alternative not available until recently. For another, present electronic technology often permits circumvention of moving parts altogether, gaining ease of assembly, greater reliability, and (one strongly suspects) economy in manufacture—even where the version with digital controls commands a premium price.

Only the toolbox—with its wrenches, pliers, screwdrivers, and the like—has retained most of its old odor: the main changes have been the proliferation of cordless, rechargeable power tools and a diversification of screwdriver tips, neither visually immediate. But the toolbox mostly remains part of the old world of fairly rigid materials and moving parts whose surfaces can shear against one another— and hence contains many grasp-and-release items.

The second trend, spanning human history, involves a reduction in the range of rotational tasks that we perform as this kind of repetitive sequence of powered arcs alternating with recovery or repositioning phases. All the common contemporary examples could be described as "grasp-and-release." For the older ones, the phrase describes the action less well as often the motions depend on the entire body rather than on just fingers, hands, or arms. What might explain this gradual abandonment of large-scale, powerful, repetitive drivers? My best guess invokes nothing beyond the gradual acquisition of more attractive alternatives. Sometimes this happened by direct substitution, as with donkey querns and cranks. Sometimes it came with full redesign of the machinery for doing a particular task, as when whims and treadmills with crown-and-lantern gears took up the burden of whole-body handspikes and the like. And some, of course, must be associated with the shift of power-demanding tasks from dependence on muscle.

Even with all those pop-off bottle tops, joy sticks, and touch-screen controls, we still do plenty of finger- and wrist-level twisting and turning, so much so and in so casual a fashion that any lurking subtleties escape notice. Consider, for instance, the matter of efficiency, our deservedly favorite measure of the quality of a mechanism. Input and output may be variously expressed, but however these are quantified, output divided by input defines an efficiency. Now think about what happens when you screw in a lightbulb. You slide its helical threads around a path set by the corresponding helical threads in the socket until the bottom of the bulb bends a tiny spring tab at the bottom of the socket. What would happen if the whole system, bulb and socket, were so well lubricated that little energy was lost (strictly, but irrelevantly, converted to heat) as you turned the bulb? Why, then every time you released your grip, the bulb would turn back the other way and undo your twist. Or else once started, it would screw itself all the way in until stopped by the spring tab at the bottom. Bottom line—the grasp-and-release system for achieving rotational motion demands inefficiency. Put another way, with efficiency defined in terms of the immediate mechanical action, effective grasp-and-release ordinarily imposes a mandatory 50 percent upper-efficiency limit. We'll get back to that "ordinarily" further along.

A second issue: nothing of any fundamental nature determines which direction tightens and which loosens, or which drives something left and which right, or which forward and which backward, and so forth. So we've established conventions, ones of such mundanity and antiquity that their origins don't seem to be matters of public record. The basic rule for turning screws, bottle tops, and the like is that clockwise tightens and counterclockwise loosens ("righty tighty, lefty loosey," as sometimes put). One can make at least a decently forceful argument for the convention based on the fact that, for a right-handed person, it yields a greater turning force for tightening than for loosening—and most of us are right-handed. More than convention comes in—one is taking advantage of some muscular asymmetry. The appendix describes a demonstration device.

Still, not all of our screws have right-hand threads—that's simply the default. For at least one common application, proper function depends on left-hand threading. A turnbuckle, as in figure 6.1a, provides a simple way to tighten a cable or pair of rods and thus bring together two elements of a framework. Turning it tightens (or loosens, depending on the direction of rotation) the connection because its two screws have opposite threads. To my eye, admittedly an experienced one, the one with the left-hand threading looks funny; more specifically, it looks as if it has steeper threads, which, of course, it doesn't. A few other mechanical devices take advantage of this mirror-image pairing, but they're not quite such everyday items. Pipe union couplings (fig. 6.1b) are close analogs of turnbuckles.

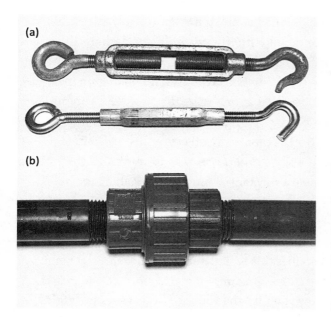

(a)

(b)

Figure 6.1. (a) A pair of turnbuckles. A similar one, mounted between two diagonal rods, keeps our rear screen door from sagging out of squareness. (b) A pipe union. The reversed threading is inside, visible only if the coupling is unscrewed.

Left-hand screws and nuts secure the left-hand pedals of bicycles, the odd threading chosen so the normal rotation will tighten rather than loosen them. As new drivers in the 1950s, many of us became aware (sometimes the hard way) that the lug nuts on the driver-side wheels of cars made by the Chrysler Corporation (before about 1960) had left-hand threading for the same reason. Ignorance of this peculiarity sometimes meant that an attempt to remove a tire left the tire more tightly fastened than ever. The worst hazard came (as I was told much more recently) when a young mechanic responded to an unwilling nut on an old car by raising the torque level on the power wrench to ever higher levels until the screw and nut broke off.

We learn the rule of screw-thread handedness at an early age, or at least we did a few years ago. I noticed that, even before his second birthday (in 1968), my son didn't hesitate when unscrewing bottle tops—even before he spoke in full sentences. And I am almost com-

pletely certain that no one had at any point specifically taught him
the convention. Some combination of watching and, more likely,
trying and self-rewarding had clearly been sufficient.

A final note on the handedness convention: in looking up material for this account, I read that at one stage the New York City sub-
way system ordered incandescent lightbulbs with left-hand threads,
along with sockets to accept them. The object of this odd shift, economical only for such a large enterprise, was to minimize theft of
bulbs, which had become a substantial problem by making the bulbs
incompatible with a standard socket. At least I hope this isn't an urban legend, however urban it may be.

An occasional gadget applies this repetitive grasp-and-release scheme
to improve the performance of something that normally manages
without it. Think of how you turn the volume knob of a radio or television (one old enough to retain knobs). The range through which
it can be rotated doesn't quite reach a single 360-degree rotation,
and, if need be, you can sweep the full range without releasing your
grip. But it affords only a crude adjustment. If some numbers and
marks are added for reference, they can't pretend particular precision. In pre-digital days, better controls, whether on the control
panel of an aircraft or on analysis equipment in a laboratory, often
came equipped with knobs with some gearing within them. They
were capable of being turned more than 360 degrees, and they came
with indicators of commensurate sophistication. The most common
kinds went by the name "ten-turn pots"—"pots" for "potentiome-
ters," coming from "potential," or voltage, which they metered out
in amounts determined by the setting one chose. A typical precision
was about 1 percent, quite good in the era of analog electronics—but
they weren't cheap.

Various gadgets are sometimes added to keep things from getting
all too literally out of hand during the release phase. If you want
to tighten a nut in a confined space, you have to repeatedly reposition your wrench, which means repeatedly taking it off the nut. A

(a)   (b)

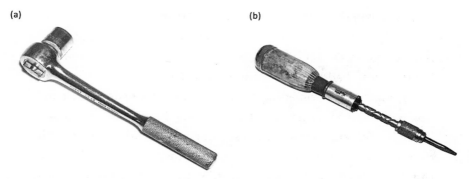

Figure 6.2. (a) An ever-useful ratchet driver with a socket attached; the lever on the back reverses the direction of ratcheting. (b) A Yankee screwdriver, still serviceable but unused since I acquired a variable-speed electric drill about forty years ago.

longtime favorite device, a ratchet driver for socket wrenches, solves the problem. Socket wrenches, for those who feel no compulsion to manipulate machinery, consist of interchangeable sockets that fit over various sized nuts and handles that fit into the sockets, extending radially so even a modest force puts a strong turning torque on the nut (fig. 6.2a). The handles of present concern (there are other types) have in their heads tiny mechanisms that lock up when turned in one direction while permitting free rotation in the other. Thus one swings the handle and turns the nut with the socket locked. But instead of having to lift the socket off the nut to regrip it at a different angle, one then just reverses the handle's motion and it freely returns while putting only trivial force on the nut. The handle goes back and forth—and the nut turns unidirectionally. A lever on the back of the handle allows the user to shift between tightening and loosening a nut. If that strikes you as an obscure mechanism, think of how one winds the spring on an old-fashioned wrist- or pocket watch. You turn the stem back and forth between thumb and forefinger. Again something inside turns unidirectionally, with the necessary slippage in the undesired direction provided by a tiny ratchet that gives perceptible clicks.

A still fancier gadget designed to evade the regripping issue has now faded into obscurity. The consummately ordinary grasp-and-release tool in any toolbox is a screwdriver. Woodworkers and boat builders who had to sink vast numbers of screws into holes commonly made use of something called a "Yankee screwdriver," a name originally derived from that of the company that made them. A Yankee screwdriver has a long, telescoping shank with a pair of helical grooves, one a left-handed helix, the other a right-handed helix (fig. 6.2b). If you push straight down on the top of the screwdriver, the shank gradually disappears into the handle while the bit with the driver turns in whichever direction you've selected—sliding along one or the other of the helical grooves. So a downward push turns the screw through several full turns; the spring-driven re-extension turns it neither way. Thus a few pushes are enough to drive the screw as far as intended. You grasp, but you needn't release at all. It happens to work less well for removing screws, but with the aid of a lockup setting on the side, it's no worse than an ordinary screwdriver. Why has this consummately clever contrivance gone out of stock in ordinary hardware stores and online catalogs? I think we need look no further than cordless, rechargeable electric screwdrivers and related non-muscularly powered alternatives—these latter have real advantages, especially of versatility and ease of use.[1]

We might view grasp-and-release devices in a sequence, beginning with those for whose operation we mobilize the fewest muscles and moving toward ones for which an ever-increasing muscle mass participates. As it happens, another sequence runs in just the opposite direction. The more delicate the task, that is, the fewer the muscles needed, the more precision we expect—and we ordinarily achieve, whether or not the extra precision is functionally critical. Those ten-turn pots mentioned earlier, and all other such control knobs, take only a pair of fingers; with a little visual feedback, they're easy to set to the nearest millimeter. The average kitchen offers many examples of devices that take only one hand to turn—the other hand most often steadies the main handle, as, for instance, with a hand-operated can opener. As expected, they achieve less precision in operation. I needn't belabor such household equipment further—except perhaps to note the nearly complete passing from the scene of equipment that you clamped to the counter with a hand-turnable screw (and thus freed that second hand). We've found that a meat grinder of this ancient design works better than at least one motorized version; unfortunately, in the present world of granite and Formica, we have trouble finding a suitable place to clamp the ancient grinder.

Most of us find shop tools less familiar than those of a kitchen. While they may do analogous tasks, the operation of these hand tools differs in one basic respect: no self-respecting shop, whether devoted to woodworking or metalworking, lacks at least one vise (not "vice"). As noted for the meat grinder, it takes the place of the second, steadying hand and, for the kinds of tasks we'll consider, it does the job much better. Conversely, no kitchen that I've ever seen has any analogous counter-mounted, general-purpose clamping tool. Nor would one be of much help unless other kitchen tools were re-designed to take advantage of its availability.

But even with something that frees the second hand, one-hand tools abound. An auger (not "augur") consists of a bit with a transverse handle at the top; it drills out medium- to large-size holes (fig. 6.3a). A gimlet is similar but has a more pointed bit suitable for softer materials (fig. 6.3b). A tapered reamer enlarges holes in sheet metal (fig. 6.3c). A hand tap cuts threads on the inside circumference of a hole (usually in metal) so a screw can be tightened in it (fig. 6.3d). And so on. While often a vise isn't strictly necessary, all do their best when the work has been properly clamped.[2]

A metal lathe, the centerpiece of every machine shop, may weigh half a ton. But the machinist operates it with an array of one-hand knobs that move with only the most modest of force—at least once the work has been clamped and the cutting tool has been brought into approximate position (fig. 6.4). While each knob has one (or sometimes two) cranks protruding, these merely speed up

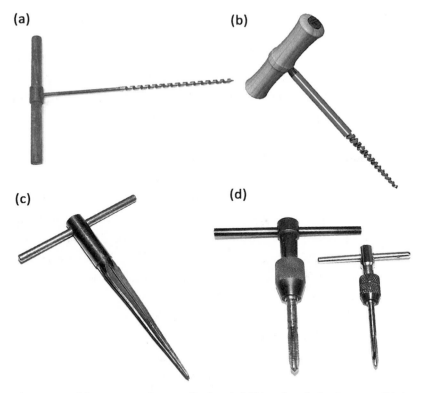

(a)

(b)

(c)

(d)

Figure 6.3. (a) An auger, this one for hand-drilling deep holes in trees. (b) A gimlet—good for small holes, starting screws, and doing duty as a corkscrew. (c) A tapered reamer for enlarging holes in metal sheet. (d) Two hand-taps, for cutting screw threads in metals and plastics.

that initial positioning. So all depends on just the same grasp-and-release, grasp-and-release. And thereby lurks a bit of trouble for every novice machinist. One needs to work slowly and deliberately—advancing the cutter too fast will cause that cutter to chatter, so the work will be gouged, or it may cause still worse evils. What matters, though, isn't the average speed of the grasp-and-release action, but the speed when grasped and turning—no jerky motions, please, please. And the nature of friction tends to promote jerkiness—it's harder to start motion than to maintain it. In formal terms, static,

Figure 6.4. A metal lathe of no particular distinction, with an enlargement of its cross-feed adjustment knob and crank.

or starting, friction, almost always exceeds dynamic, or moving, friction—as we made much of in chapter 3. (Of course, repositioning the hand when not advancing the tool—that is, during the release phase—can be done as fast as one wishes.) Lathes will reappear in later chapters; they may not be the most common tools in home workshops, but they have played a central role in our mechanical technology for centuries.

What about small two-hand grasp-and-release tools? In their larger incarnations, all the tools that have already appeared need a second hand. To these we can add a few tools that almost always require two hands. For some, hands grasp opposite sides of the tool, that is, 180 degrees apart. The mate to the hand tap is the die holder (fig. 6.5), generally two-armed and symmetrical. Socket wrenches sometimes have paired handles (or handles that can be slid into a symmetrical position) for stronger and more balanced application of turning force. With many tools, both hands grasp the same side,

Figure 6.5. Two die holders with dies inserted. These cut outer threads, those on screws, as opposed to the inner threads of sockets, holes, and nuts.

and the whole body can be mobilized for pushing or pulling—only the length of the lever arm, the handle, limits the potential turning force. The biggest such tool in everyday use may be a pipe wrench; you should be impressed with the size of wrench that can be purchased in an ordinary hardware store or tool department. Old and well-rusted steel pipes often take serious persuading to separate from their couplings. The last time I looked, a pipe wrench with a two-foot-long handle was a stock item. So we've moved well beyond grasp-and-release tools to which the word "small" might be appropriate.

I have the impression, though, that the largest of these tools, mainly wrenches, have receded into history. In part, we may be looking at yet another instance of the spread of power sources other than muscle, sources themselves rotational or readily made to rotate. In part, we now prefer designs that use many smaller screws and nuts rather than one or a few large ones—that way we can distribute forces more uniformly and gain the safety that comes with redundancy. For a close-to-home example, count the lug nuts on a wheel of your car; it probably has five. Then count the nuts on the front wheel of a large truck—yes, they're bigger, but they're much more numerous rather than being enlarged in proportion to the size of the vehicle. Still, large wrenches do persist here and there. The largest one I ever handled tightened the top of a so-called bomb calorimeter, squeezing it down against a lead gasket.[3] As I recall, the wrench extended outward five or six feet, with the calorimeter held snugly in a special mount attached to the wall of the building. That gasket ensured a seal that could withstand the high pressure of the gases released after full combustion of the contents of the already-pressurized vessel. The calorimeter was sunk in a water bath; detonation was electrical, the explosion externally audible as a dull thud. The energy released warmed the bath, which one measured; the rise in temperature and the volume of the water established the amount of heat released.

Back to the last entry in the initial list, the steering wheel of a car, without doubt a repetitive grasp-and-release tool. How we manipulate steering wheels (nicely literal word, "manipulate" in this context!) gives little hint of their history. As you might guess, steering wheels descend from ships' wheels, the traditional steering controls of all but very small vessels, whether wind-powered or motorized. While ocean-crossing boats in both the Atlantic and Pacific basins go back about a millennium, they acquired these vertical wheels at their helms only about three hundred years ago. Rear rudders or rudder paddles go back much further—at least to the ancient Egyptians. Some were mounted on one side, some were paired and, in not-so-ancient cultures, some were mounted on the rear transom as they are on small boats today. It has been suggested that the "port" side of a boat was originally the side without the rudder, the side that could contact rock or dock without damage to any movable components.

Again going back to the Egyptians, a forward-directed tiller at deck level might be attached to the rudder to help the mariner swing it to one side or the other. Push the tiller to the right, and the boat turns to the left, something familiar to anyone who has sailed a small boat. As ships grew larger—to the hundreds of tons of the massive warships of the Spanish Armada, for instance—tillers acquired additional devices to gain mechanical advantage.[4] The task was not trivial when turning a large vessel with the wind in its sails. We have good evidence (from a recovered wreck) that by 1703 the English navy had begun using ships' wheels instead of augmented tillers.[5] Ropes and pulleys led astern from the wheel, whose forward location afforded the helmsman a far better view of what might lie ahead than had been possible earlier. In effect, ropes and pulleys decoupled the relative locations of wheel and rudder. From the start, ships' wheels appear to have been equipped with their iconic set of radial handgrips. Indeed, it's hard to imagine managing the wheel of a large vessel without them while standing on a heaving deck in a high wind and perhaps an icy rain.

Now nothing fixes which direction of wheel rotation corresponds to what change of course, and the convention for the earliest ships' wheels had clockwise rotation—to the right on the wheel's top—turning the ship to the left, just as if the helmsman had hold of a tiller. We'd find that unnatural, and apparently the unnaturalness reflects more than our personal upbringings, since the convention was soon changed to the one we know, the one that has been universal ever since for all craft, whether at sea, on land, or in the air. Turn the top of the wheel to the right, and the craft turns right.

Steering wheels on automobiles, like those on ships, replaced tillers, with the change happening two hundred years later, mainly in the first decade of the twentieth century. Here the direction convention posed no problem inasmuch as the tiller, controlling the front wheels, extended rearward rather than forward—push the tiller so it points to the right and the car turns right. For cars, gaining better visibility meant little, but the other advantage of the wheel was more compelling. The wheel, whether on car or ship, can be linked up to give any mechanical advantage the designer wishes. In particular, it can allow a mere human to control a heavy vehicle by incorporating gears in its linkage that require that it turn many revolutions within its full range. In doing so, it also gives the operator fine control when that's what is needed, because a long total travel ensures that minor adjustments can be done with due deliberation and that slight twitches matter little. Even so, we easily forget how much force it takes to turn the steering wheel of a car now that all cars come equipped with some form of engine-derived steering wheel assistance. In my youth, one was all too aware of the work of parallel parking in a minimal space, or of driving with under-inflated tires.[6] Bus and truck drivers didn't even try to turn the large steering wheels of their vehicles unless those vehicles were moving forward or backward.

Even if we no longer need the force advantage, we do love our steering wheels. Every so often some substitute makes the magazine rack, but none ever goes into mass production. We even equip vehicles with them that could, with no loss of functionality, be more simply built without them, vehicles such as go-carts, riding lawn mowers, and the like. For these latter, hinged, telescoping tillers would be so simple and so versatile—and so redolent with public evidence of frugality.

Consider again those lovely varnished handgrips that extend outward from every proper ship's wheel. In a sense, those grips make the wheel itself superfluous. Just extend the grips inward to some central hub and ask what, beyond tradition and aesthetics, has been lost? What we have thereby devised is what we will call—using a term in an unusually general sense—"handspikes," as in figure 6.6a. These are old—too old and too simple for us to have any good sense of when they first came into use. They represent only minor elaborations of levers. For instance, if (as in fig. 6.6b) you roll a wide log by repeatedly sticking a pole in the ground beneath and behind it and then lifting and repositioning that pole, you have applied something equally well described as a lever or (in the present general sense) a handspike. Beyond this simplicity, their antiquity quite likely bears some relationship with the relative non-antiquity of proper cranks, the subject of another chapter. Hubs with cranks and hubs with handspikes not only do much the same things but can often be interchanged with little redesign of their associated machinery. In other words, they accomplish nearly the same thing in terms of relative turning force, power, and so forth.

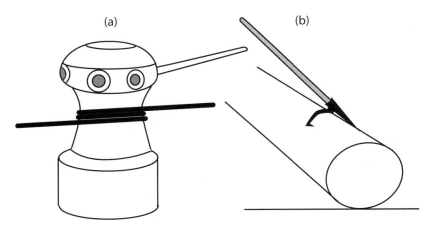

***Figure* 6.6.** (a) An anchor-raising capstan is turned by inserting a handspike into successive sockets. (b) A peavey eases the task of rolling a heavy log.

We currently bring handspikes to bear for just a few of our pressing tasks, applications in which we need to apply force smoothly or to smoothly increase the force at the end of the travel of something like a vise or clamp. The most familiar handspikes grace the right side of the top of every drill press, as in figure 6.7. These are bench-mounted or floor-standing power drills in which turning a handspike (or several, sequentially) lowers a turning drill toward and then into the object receiving its attention. They usually have three or four radial handspikes, so the chuck can first be lowered rapidly and then one or another of the handspikes will be conveniently placed and angled for the final, forceful, and carefully applied movement into the work. Some vises—but not the most common varieties—have similar multiple-spike shafts, as do a variety of other hand-operated industrial presses. In none of them is the operator called on to apply especially great force, and none asks for sustained power output.

That restriction of handspikes to low forces, though, was not the case before the advent of easy motorization. Handspikes once served as serious power-transmission devices, filling various niches where treadmills and waterwheels would have been unnecessarily

Figure 6.7. A drill press of the most ordinary sort, with its permanent handspikes, which serve to lower the chuck and contained drill bit into the workpiece.

powerful—and thus expensive and inefficient. In this service, their history resembles that of cranks—once large and powerful but now almost entirely relegated to very low-force, low-power applications. For that matter, shafts with handspikes and shafts with cranks appear to be largely interchangeable in the fifteenth- and sixteenth-century books of mechanical contrivances to which we keep returning. Agricola shows several shafts that form parts of barrel lifters and pumps with handspikes at one end and large cranks at the other, as if to tell the potential builder that either will suffice.[7] Again, these tasks asked for real power; as Agricola put it, ". . . the levers by two men, of whom one pulls while the other pushes; all windlass workers, whatever kind of a machine they may turn, are necessarily robust that they can sustain such great toil."[8]

In fact, all of our favorite Renaissance sources make substantial use of shafts with handspikes; figure 6.8 gives some examples. The

Figure 6.8. (a) Agricola's well-hoist with handspikes on a horizontally mounted capstan; (b) Strada's chain pump. Both are two-person handspike devices.

number of wheels varies from one to four, the number of operators per wheel is sometimes one, sometimes two, and the wheels have between four and eight spikes. They do service for all manner of pumps, presses, lifts, dredges, moat-bridging machinery (for sieges), and even a crossbow-like artillery piece. These last two come from Ramelli, professionally a military engineer. Curiously, Ramelli only puts cranks on his civilian appliances, reserving the handspikes for his military gear, which include far more than just weaponry.

Going back further in time increases our uncertainties—we become all too dependent on interpretation of words without illustrations, on guesses as to how static artifacts moved, and on the form of parts that decomposed long ago. As mentioned two chapters ago, small querns of classical antiquity sometimes had rings of radial holes in their top stones, rings into which wooden extensions—handspikes—would have fitted quite nicely. But did the person operating the quern simply grasp a pair opposite each other and move the top stone back and forth, with no release phase, or did these handspikes signal a grasp-and-release mechanism of the form we've been considering in this chapter? I'd tentatively opt for the latter as the more probable on the admittedly shaky grounds that back-and-forth would have needed only a single pair of handspikes, and making holes in stones could not have been a casual activity.

Without a doubt, the ultimate in power-absorbing grasp-and-release devices were the great torsion weapons of classical antiquity—the so-called ballistae in which the artillerymen twisted bundles of bovine tendon. We know more about ballistae from primary sources than about most of the mechanical technology of their time: much has been written about them in the past century or so, and various people have experimented with models to evaluate their performance.[9] What rarely attracts notice is the self-evident but critical fact that these were entirely powered by human muscle—none of that "better killing through chemistry" (to pervert a recent corporate slogan). They never approached the capabilities of the trebuchets of the past millennium or so, in part because their design, by contrast with the latter, precluded simultaneous efforts by more than two artillerymen. At the same time, and as we'll see, they managed to throw remarkably large masses remarkably long distances with remarkable efficiency. Still, as muscle-powered machines, these torsion weapons had to be severely power-limited. We'll return to torsion weapons shortly, reversing the historical sequence to look at a sequence of increasing mechanical complexity instead.

Say you want to heave a heavy projectile some decent distance with military or otherwise malevolent mission. With clever application of levers and ropes, you might couple a large number of humans to the task, commanding the multitude with a signal (Joshua's horn at Jericho, for instance) to pull simultaneously. This real-time exertion powered the earliest trebuchets, so-called traction trebuchets—as in figure 6.9a—which somehow applied the synchronized efforts of hundreds of men.[10] The sum of the power put out by each man, less whatever frictional and other losses could not be avoided, equals the power given the projectile. That power, then, equals its mass times

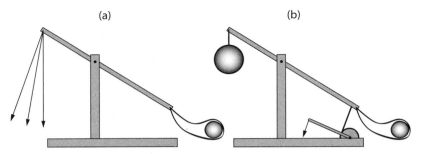

Figure 6.9. Put in the simplest diagrammatic form, the contrast between traction (a) and counterweight (b) trebuchets. Power input to the former is provided by a number of men pulling on ropes (represented here by arrows) and the energy input is immediately conveyed to the projectile. In contrast, power input to the counterweight trebuchet (b) is via a lever or handspikes that turn a capstan; the energy is stored in the raised weight to the left and imparted to the projectile when the longer lever to the right is released.

the speed at which it's launched. Straightforward so far—power put in by the artillerymen (less losses) equals power put into the missile.

The trouble with a traction trebuchet is that too much effort gives too little yield—too many people, too little power output. But the underlying physics permits a fine trick, one that violates no law nor asks that muscle do anything better. What's needed, and asks only quite lawful physics, is a power amplifier, some scheme that allows a low level of power to be put in over a relatively long period of time while a high level of power comes out over a short period. That's what an archer does, slowly drawing back the bow and then allowing the stored power to be released rapidly—archery gains its intrinsic superiority over spear throwing by using power amplification. A fundamental principle of physics, the conservation of energy, declares that the work put out can't exceed the work put in. But it implies conservation of power only in the absence of power amplification—put work in slowly, take work out rapidly, all quite legally.

Amplifying power, though, requires an additional element, specifically a way to store the work (or energy) put in. Of the many potential systems, two were practical for pre-industrial technologies. The great counterweight trebuchets of the Middle Ages threw far more mass much farther than did the traction trebuchets developed earlier. As in figure 6.9b, they stored work gravitationally by slowly hoisting a weight, reportedly as much as 10 tons. Typically several people turning handspikes did the heavy lifting. Release of some catch then permitted the weight to fall back down and, in falling, to heave the projectile. They were far from efficient, with lots of the stored energy wasted in the final impact of the weight with the ground. But they were big, simple, and, as military hardware goes, cheap. Weight, as dirt or stone, would always be available on site.

And we know a great deal about them. Good descriptions and illustrations survive, and they lend themselves to all manner of re-constructions and model building—no materials are either difficult to obtain or subject to rapid deterioration. Many of the models use cranks to pull up the counterweights, something originally rare. An occasional reconstruction employs a treadwheel, as did some of the originals. The web abounds in pictures; figure 6.10 is a typical recon-struction, this one minus any of the rigging and thus with its coun-terweight downward, as we saw it on the grounds of Dover Castle, in England, about a decade ago.

Back to torsion weapons. The other way work (or energy) could be stored was elastically. In archery, the work of drawing bends a bow further out of its preferred shape, and elastic recoil then restores its earlier shape while sending the arrow on its way. Similarly, torsion artillery depended on elastic storage for power amplification. But instead of bending wooden beams, it stretched animal protein, pref-erably a protein called collagen in the form of bovine leg tendon or, when that was not at hand, keratin, mainly as ropes of human hair. Collagen (most of a tendon) stores energy very well, absorbing lots of energy relative to its weight, and it releases the energy efficiently, giving back over 90 percent of what was put in. Still, however good it might be as an elastic, it can't be treated as if it were a rub-ber band—it stretches by only about 10 percent before permanent

Figure 6.10. (*Left*) Dover Castle's trebuchet, with tourist (= spouse) for scale, plus the ratchet mechanism (*upper right*) that held the arm and sling cocked (and made handspikes practical) and the box into which ballasting stones could be loaded on site (*lower right*).

damage occurs. So the work—force times distance—must be put in as huge force but very little stretching distance. The ancients cleverly did this by twisting tendons, bundled together in a way no longer fully understood.

In a ballista[11] (fig. 6.11), two bundles of tendon extended between the sturdiest of wooden supports, with metal pieces top and bottom, where stresses were greatest. Each bundle received the end of a lightweight wooden rod, with the opposite ends of the rods linked by the bowstring. With lightweight wooden rods moving more slowly than the load and losing most of their speed by the end of

Figure 6.11. A model ballista I built for demonstration in lectures; the projectile, safely shootable at an audience, is a practice golf ball (or Ping-Pong or Styrofoam balls). In reality, the handspikes should have been much larger. Construction details and dimensions are given in the appendix.

their travel, little energy was lost—these weapons must have been quite efficient. Remember that we're looking at large-scale weaponry powered entirely by human muscle, so their efficiency was a major component of practical efficacy. Efficiency also mattered in a secondary way—more work going into the projectile meant less work generating troublesome recoil of the machine.

Almost all drawings of ballistae show them equipped with hand-spikes, but to my biomechanically prejudiced eyes, the relatively short handspikes usually figured look unequal to the task, the impetus for all these words as well as for the projectiles. It does appear as if a ballista could accept no more (and probably used no fewer) than two artillerymen, so those people—or relays of people—had to suffice. What did the task imposed by a large ballista demand of them? I think a reasonably reliable estimate can be made from a productive collaboration between a scholar of classical technology, J. G. Landels, and a particularly insightful mechanical engineer, J. E. Gordon, both of Reading University (UK).[12] They note that large numbers of 90-pound (40-kilogram) stone spheres lie beneath Carthage harbor, in North Africa, debris from the Roman siege, over two millennia ago, that ultimately destroyed Carthage. They also note that big ballistae could shoot about 400 meters, far enough to keep the valued artillerymen beyond the range of opposing archers. We can assume from contemporary data that a well-trained man in good condition should be able to sustain about 200 watts output. If we assume roughly 50 percent efficiency for the overall machine and two artillerymen, then some straightforward calculations, a good homework problem for a first-year physics class, suggests a rate of firing of one shot every 7 minutes.

Now that's working the ballista's operators at about the maximum level that we can expect from humans working on all but the very best of our human-machine couplings. They might have put out slightly more work if pedaling racing bicycles or human-powered aircraft or if they were rowing racing shells. In any case, short (literal) handspikes would preclude adequate energy transfer—for aerobically taxing work, we have to mobilize the lower body and leg muscles. So I will go out on a limb and suggest that, yes, grasp-and-release spikes did the job, but on the largest of the ballistae, these amounted to something closer to whole-body spikes. Each of the paired artillerymen would have had a stiff pole of almost his body height; this would be inserted into a hole in the front of the capstan of the ballista and then pulled backward, turning the capstan. At the same time, the other artilleryman would have positioned his stiff pole to engage a similar hole in the front of the other end of the capstan so he could take his turn pulling backward and turning the capstan. Good coordination would have taken practice, and the holes might have needed entrance cones for guidance, but the system sounds simple and reliable.

Another approach also suggests that this hypothetical alternate body-spike system yields more reasonable numbers than any smaller handspikes. A human, pulling horizontally, as one might on a vertical body-spike of roughly one's height, can exert a force of about 100 pounds. That means the minimum pulling distance to wind in enough energy for the shot (40 kilograms going 400 meters at 50 percent efficiency) has to be about 1,000 feet. Assuming a reasonable pulling radius, the capstan thus has to be turned about thirty times per shot. For a shot every 7 minutes, that's a turn every 12 seconds or about five full turns per minute, reasonable enough given sufficient practice to develop the requisite teamwork. Furthermore, the pole-in-socket system puts the capstan on the ground or on the ship's deck. Even with minimal recoil, that's a more likely location for the driver for 90-pound projectiles than having the capstan at waist or shoulder height, as in most illustrations.[13] At least one figure does show the capstan on the ground, which would preclude the usual ring of fixed handspikes. That's E. W. Marsden's interpretation of the brief descriptive material in the tenth book of the Roman Vitruvius—who explicitly considered ballistae capable of throwing very large stones.[14]

Finally—if perhaps parenthetically—why did these splendidly efficient machines give way to the much less efficient counterweight trebuchets? Most likely the ballistae had an Achilles' heel of a kind that's too literal to pass for a mere pun—bovine leg tendon doesn't grow on trees. We have good figures for the relevant datum for tendon, something called "strain energy storage." So we can easily figure out how much tendon would be needed to store up the energy needed to heave a 90-pound stone 400 meters at 50 percent efficiency—it comes to over 60 pounds. I don't know how much can be harvested from each cow, bull, or ox, but the yield must surely be far less than that. And, quite as surely, the material is both perishable and subject to irreversible damage from overstraining. Trebuchets have to be far easier to build and maintain; most importantly, they can be scaled up to sizes quite unimaginable if one has to procure ever-greater supplies of tendon and tolerate ever-decreasing rates of launching (one probably shouldn't say "firing") shots.

A few other grasp-and-release devices need little elaboration but shouldn't go unmentioned. The third chapter touched on potter's wheels that were driven by kick wheels—these aren't grasped in exactly the same way, but the principle is the same. And a wheel on a horizontal shaft can be turned hand over hand if a loose rope dangles from a pulley. Applying your two hands alternately means that a ratchet to prevent retrograde motion during a release phase becomes unnecessary—although it limits the power you can apply to the task. You might hoist a bucket from a well this way—although we humans, as befits members of an order of primitively arboreal and thus climbing mammals, much prefer to pull downward rather than upward.

[7]

Turning and Unturning

We turn to still another way to trick our non-rotating muscle motors into acting as rotational drivers. While the devices using the trick aren't as common among our household gear as grasp-and-release gadgets, they're far from obscure. These are rotational devices that we turn without relinquishing contact—but ones that we (or some return system) then turn by the same amount in the opposite direction. As a result, they perform some rotational task, truly rotating in doing so; nevertheless, on average they do so without rotating at all. What are we talking about? I can introduce them by reaching for the one closest at hand, just opening the top drawer to the immediate right of my desk. From it I take a metal tape measure of the type that uncoils and emerges from its case when you pull its end. When you release it by tripping a catch (a sliding switch on mine), the extended tape then disappears back into the case, returning to its original coiled form as it's pulled back by a spiral spring inside. So the tape obligingly unwinds and rewinds as you find convenient.

Examples of gadgets that depend on this particular trick, turning in one direction of rotation and then turning by the same amount in the other, don't just jump out at you as did the grasp-and-release devices. Still, they're common enough, and, as we'll see, they can boast of respectable antiquity and cultural range. A simple list impresses one with their diversity but reveals little about any constraints on their design and utility. Better, I think, is to divide them based on the simplest of questions. How does each do what we might call the return or reverse rotation? The tape measure, for instance, made use of that internal spiral spring, strained as you pulled the tape out, and then, in its eagerness to relieve the strain, more than willing to pull the tape back in.

So what are the possibilities?

- The return might, like the initial or power turn, be driven by *direct muscular action*. It might not be distinguishable as a return stroke in any properly functional sense—just a stroke in the opposite direction.
- The return might be driven by a mechanical spring, as in that tape measure. The essence of what's happening here is *elastic energy storage*—the operator powers both strokes, simply doing extra work during what we might call the volitional stroke that is subsequently used to drive the return stroke.
- The return might be driven by the descent of a weight that had been hoisted during the stroke in which the operator was exerting muscular effort. Here we have *gravitational energy storage*, as in the giant counterweight trebuchets of the last chapter.
- The return might be driven by a flywheel or some equivalent. In effect, this *inertial energy storage* uses rotational kinetic energy, the persistent spin of some relatively heavy disk, energy that can be reinvested in a return stroke.

Not that these exhaust the range of physical possibilities. One can envision storing energy for return strokes through rechargeable batteries (or electrical capacitors), compressing gases by pushing pistons into blind-ended cylinders and so forth. But I don't have obvious examples at hand among our practical artifacts, past or present

Figure 7.1. The mini-malist's knife sharpener. The device has only one moving part; the ceramic abrasive disk is in the middle. This one is about three inches in both length and diameter.

or even suggestions for utilitarian items that you might find amusing to build.

My favorite example of a purely muscular push-pull gadget may not be as familiar as a retractable tape measure, but I do think no home should be without one—in my experiences most of the alternatives either cost more or don't work as well. Unfortunately, they're not as common in stores as they were a few decades ago. A rolling wheel knife sharpener has, depending on how you think about these things, either no moving parts or only one. At least no part of it, as in fig-ure 7.1, moves with respect to any other of its parts. One just draws the knife back and forth, laying its flat surface against that of one plastic wheel and its blade against the sharpening stone; one then does the same thing with the other side of the blade and the other plastic wheel. The longer the knife, the more revolutions it asks of the wheel, but that's of no functional consequence, worth mention-ing only on account of the subject of the present book.

That sharpener strikes me as a particularly clever kind of back-

and-forth roller. These rollers are really quite ordinary. Think of rolling pins of the kind with handles on their ends, of paint rollers, of lint-removing rollers, of some muscle-massaging rollers, and of any others that might occur to you and don't happen to occur to me as I write this. Their common characteristic, along with the knife sharpener, is that none needs exact matching of the amount of rotation in the two directions. That seems to be a common characteristic (and slight advantage) of things directly driven by muscle in both directions. But, unlike the knife sharpener, the other examples require a set of bearings, even if the demands on the bearings are modest and easily satisfied.

Not only is zero net rotation by purely muscular back-and-forth motion simple, it's also the most ancient technique of present relevance. It forms the basis of the bow drill in particular, an early and widely used tool. Evidence of its commonness, admittedly indirect, goes back not just before recorded history but before the agricultural revolution—to the upper Paleolithic, very roughly 20,000 years ago.[1] If you do a search starting with "bow drill," almost all the hits refer to the way it can be employed to start a fire. More interesting and perhaps older are the various drilling tasks done by this not inappropriately named device. Figure 7.2 shows its main components—a shaft or spindle with some kind of bit, a bearing block or hand piece, and a bow with a loose bowstring that can loop around the shaft. Sawing back and forth with the bow turns the shaft first in one direction and then the other. The operator's other hand holds the bearing block, which presses directly down the axis of the shaft, making it very easy to keep the shaft from wobbling as well as to control the downward force. I find it easier to keep a bow drill aligned than either the contemporary brace-and-bit or the egg-beater type of hand drill—it asks only a negligible level of skill. With a modern bit, one stroke becomes a wasted recovery one, but nothing precludes adopting a bit that bores when rotated in either direction. In short, this is one terrific tool.

Lost in the mists of time is the origin—or origins—of the bow

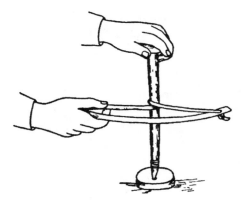

Figure 7.2. A classic bow drill—easy to make and easy to use for either drilling or fire starting (from McGuire, 1896).

drill. Some have speculated that it either followed from the development of archery or perhaps aided in that development. And two other drilling arrangements that appear functionally related might or might not be related historically. If one rubs one's palms back and forth against each other with the long cylindrical shaft of a drill in between them, then that shaft will turn as a so-called hand drill, first one way and then the other (as in fig. 7.3a), just as it does in a bow drill. The hand drill provides no obvious equivalent of the downward thrust of the bearing block and tempts dismissal as ineffective. That, though, may in part represent the unconsidered prejudice of us moderns accustomed to commercial wooden dowels of unvarying diameter. Rubbing a sufficiently bumpy cylinder should provide considerable opportunity to apply axial thrust. Alternatively, one can run a strap around the shaft (without a bow) and turn the drill by pulling the strap alternately left and right while holding the shaft in position by mouth or with the aid of an assistant (fig. 7.3b). For the strap drill, the necessary top support automatically provides downward thrust.

What has long been clear is that bow drills are not only very old but remarkably ubiquitous. An ancient culture that lacked them attracts more notice than one that does. Except for a few odd cases, what got drilled was whatever hard material functioned better if perforated by small holes—bone, stone, ceramics, dry wood, teeth.

(a)   **(b)**

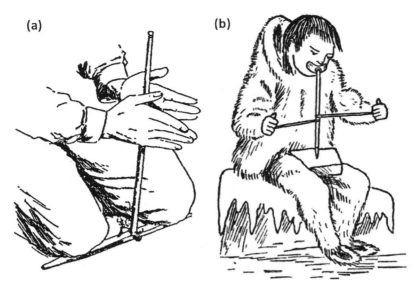

Figure 7.3. (a) A hand drill, this one with a second stick as a fire starter, as done by Indians of northeastern North America (Hough, 1890). (b) A strap drill, with a mouthpiece that's quite a lot fancier than the diagram implies, as demonstrated by an Inuit (from McGuire, 1896).

For some reason, humans have long been fond of beads, whether for adornment, as indicators of wealth, or, as with the wampum of native North Americans, an actual medium of exchange. Beads can be strung only if punctured, which almost always implies some kind of drilling. And that was apparently a major role of bow drills in many cultures over many millennia. Drilling tiny holes in pebbles of most kinds of stone does not happen easily, but a few of the softer varieties take kindly to the process, as do various ceramics. Particularly good are the shells of marine mollusks and the dried bones of large vertebrates, both with some internal fibrous structure and a little proteinaceous binder. I once found a bag of scallop shells that was about to be discarded; drilling a tiny hole near the thin edge of each with ordinary tools proved no trick at all. To my surprise not one of these thin shells shattered or even cracked in the process.[2]

What underlies this odd advantage of biologically derived materials is that they differ from almost all non-biological hard stuff in being tougher—more resistant to crack propagation and hence more forgiving when drilled. Even the enamel of your teeth, the hardest material in your body, has a little toughness-increasing protein (whose burning makes the sulfurous smell when the dentist's drill attacks it) included. To put some numbers on the variable, a typical value for the fracture toughness of a stone might be 20 (the units, of little importance here, are joules per square meter). Tooth enamel is about 200 and tooth dentine about 550. Mollusk shell runs around 1,000, and cow leg bone around 1,700.[3]

So teeth take to drilling, and the practice goes back much further than you might imagine—we'll put up with a lot when faced with the agony of a toothache. Skeletons have been recovered in Pakistan with drilled crowns of their molar teeth, drillings clearly done while the individuals were alive—between 7,500 and 9,000 years ago! Flint drill heads were recovered as well, and work with models indicates a high probability that small bow drills did the job speedily and effectively.[4] Not that the culture limited its drilling to dentistry—the same site yielded beads of many materials, including turquoise and lapis lazuli. Some human tastes don't change a lot with time. And others beside these particular Neolithic folk practiced dental drilling. Since skulls with teeth preserve well, especially in dry climates, we know that quite a lot of dental drilling was done in classical Mediterranean cultures, in particular by Egyptians, Phoenicians, and Etruscans. In some cases, wires held prostheses in place by wrapping around teeth and even threading through holes in them.

Besides depending on wood fitted with stone and crystal cutting tips, bow-drill shafts could be made of copper, bronze, or iron. Sometimes their cutting depended mainly on abrasive powder poured, sometimes with an oily carrier, into the cutting interface. Or a shaft could be fitted with a hollow, cylindrical outer portion near its lower end, the latter filled with sand or other abrasive. Turning the shaft then fed sand into an annular cut; carrying this cut through to the

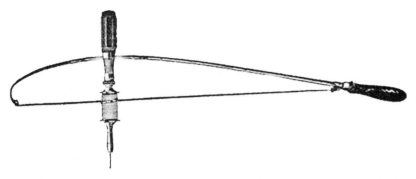

Figure 7.4. A late nineteenth-century bow drill for small-scale work. I doubt if the more recent eggbeater type of hand drill could achieve its smoothness of operation and freedom from bit wobble.

other side of the work produced a larger hole and a cutout plug. Beginning in about the fourteenth century, bow-drill drivers also turned horizontal lathes, about which we'll have more to say in a few pages.

If bow drills were such excellent tools in terms of stability and versatility, why did they pass from the scene? The simple answer is that they actually lasted until quite recently, knocked out only by the egg-beater type of hand drill, which had the advantage of compactness for carrying in a portable toolbox, and the portable electric drill, which took advantage of the near-universal availability of on-demand electrical power. (A brace-and-bit turns more slowly, so an ordinary electric drill doesn't do quite the same job.) Through the end of the nineteenth century, fine bow drills appeared in the catalogs of tool companies (as in fig. 7.4), and drill shaft design still generated an occasional patent. Somehow we've turned our back on what to me seems a swell feature not easily achieved with their successors. Changing the diameter of the part of the shaft about which the cord wraps changes the speed of rotation of the bit relative to the operator's stroking rate, and the change needn't require stopping the operation.

Not only were bow drills still extant during the nineteenth century, but the basic scheme for driving a shaft acquired an apparently novel application around the middle of the century. If you drop seeds onto a disk that spins horizontally—that is, one with a vertical shaft—they will be flung in a wide circle. You can buy (and I owned one for many years) a seed spreader for lawns that held both seed and fertilizer in a hopper above such a disk. Spinning happened as you pushed it along, courtesy of the rotation of one of its wheels and a pair of bevel gears. Its immediate ancestor, a normal item on many small farms, was carried via a neck strap by the operator, and a hand crank turned the disk—wheels suitable for lawns would be completely dysfunctional on plowed and harrowed fields. The ancestor of that ancestor had no crank and thus avoided any bevel gears, which ask for some effort to keep aligned and unclogged. Instead, just below the spinning disk, as in figure 7.5, a loop of a leather thong from a bow wrapped around an extension of the vertical shaft. The seeder moved the bow back and forth while walking along, spinning the shaft and flinging the seed. Such a fiddle-bow seeder could broadcast seed over about two acres an hour and must have given better uniformity than hand casting.[5]

Note that, with a sufficiently loose string, the dexterous operator (bowyer?) can make a shaft of a bow drill or any of its derivatives

(a)   (b)

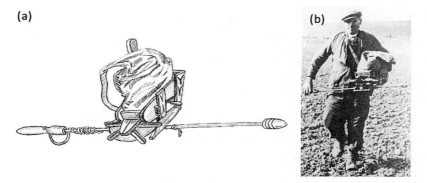

Figure 7.5. A so-called seed fiddle or fiddle-bow seeder, used during the latter half of the nineteenth and early in the twentieth century on small farms, plus an old photograph. (See Antique Farm Tools: http://www.antiquefarmtools .info/page3.htm.)

rotate unidirectionally. All that's needed is a shaft with enough angular momentum, perhaps one equipped with a flywheel. The bow
is moved away from the shaft sufficiently to tighten the loop around
the shaft during the power stroke. Then it's moved closer to the shaft
during the recovery stroke so the loop loosens and the shaft continues spinning on its own. While the trick must have been familiar to
many ancient drillers, the extent to which the arrangement saw use
remains unknown. A friend reports seeing what may have been this
device in India, with a short bow, an elliptical motion, and a flywheel
to keep the shaft moving.[6]

Back to that spring-loaded tape measure. In a sense, the thing
stretches the point just a bit for inclusion here in that what rotation
does take place can't be considered functionally central. It just allows
you to roll the tape up for convenient storage. Still, the coil of tape
rolls one way to extend, the other way to retract; you pull on it to
extend it, while its internal spiral spring pulls on its other end to retract it. That's the case for other spring-loaded extending devices. An
occasional hotel bathroom has an extending clothes line that works
the same way, as do seat belt retractors in automobiles,[7] and pedestrian access-blocking tapes at the entrances to various public venues.
On a larger scale, some manual garage-door openers make a spiral
spring serve a slightly different purpose, providing an offset for the
door's weight, but working in more or less the same fashion. It's a
handy scheme for a technology equipped with inexpensive, reliable
springs of the right geometry.

Can we identify devices for which rotation itself plays the central
role? I'd suggest the way starter cords work on most lawn mowers,
chain saws, gasoline-powered blowers, and the like, as in figure 7.6.
You pull the cord and, with luck (have you remembered to set the
choke or pressed the priming pump, perhaps?), the engine fires up.
The cord retracts on its own, and if the motor fired up, off you go.
If not, you pull the retracted cord once again, perhaps adding well-
chosen epithets of approbation or invective. The engine turns only

Figure 7.6. The hand-starter of a basic lawn mower. A spring (or springs) causes the cord to rewind, uncoupled from the now-turning motor. Fixing such assemblies can be tricky—the springs are high-strung and malevolent.

one way; the spring-powered rotation that accompanies cord retraction bypasses the engine by means of a simple slip clutch or ratchet, not all that different from the one on the ratchet-equipped socket wrench that was introduced in the last chapter. Early gasoline-powered lawn mowers required that the operator manually wind the starter cord prior to each attempt to start the engine; I assure the reader that the spring-powered return represents a great advance in every way except as instigator of creative vituperation.

Even here, history lurks. For a thousand years, spring-return lathes have graced just about every proper woodworking shop—the term "wood turning" referred to lathe work. Since we haven't yet run into lathes in this account and since I suspect that most readers have never experienced the pleasure of making something on one, a few preliminary words of explanation—and then early history—ought to precede any talk about ones with spring-return mechanisms.

Ordinarily, if you drill a hole in some piece of material, the drill turns while the material (the "work") remains stationary. A lathe reverses this—the work turns while the drill or other cutting tool is slowly advanced into the work without turning. In this sense, a potter's wheel can be (and often is) considered a form of lathe. The earliest lathes may possibly have turned the work about a vertical axis, but the much more usual thing would have been lathes in which the work spins horizontally. The horizontal arrangement is simply more convenient, whether one considers driving force, arrangement of bearings, or overall versatility.

As a practical matter, we distinguish between wood lathes and metal lathes, since either type adapts poorly to use with the other material. Critical in both is the rigid support of the turning work. If the work is short, grasping it in some single chuck (headstock) may be sufficient; if the work extends lengthwise, then support at or near both ends must be provided, as in figure 7.7. Wood lathes may go back to classical Celtic and Etruscan cultures and perhaps even Bronze Age Mesopotamia, but metal lathes have been around for only a few centuries. Sources vary in their views on lathe origins, although there does seem to be general agreement that ancient Egypt made little or no use of any lathes, despite excellent and extensive woodworking.[8]

So how to drive a wood lathe? Two adaptations of drilling techniques saw early, if limited, applications. If a strap is wound around the work or some extension of it, an assistant to the woodworker can draw the strap back and forth, with one hand on each end of the strap. The ancient Egyptians appear to have turned wooden objects this way, although not especially early. Or a bow can connect the two ends of the strap, just as in a bow drill. Now the woodworker can provide the power, but doing so requires use of one hand. Both techniques represent muscular push-pull mechanisms of the kind we talked about a few pages back. As with bow drills, a power stroke alternates with a recovery stroke during which the cutting tool is out of action.

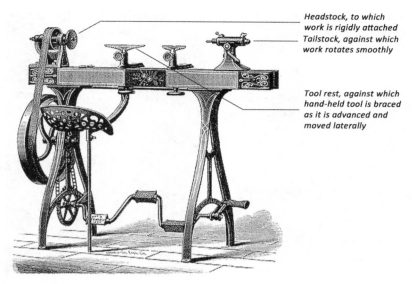

Figure 7.7. A wood lathe, this one a #3 Velocipede, from the 1885 catalog of W. F. and John Barnes Company, muscle driven as appropriate for this book, but otherwise not basically different from present hobbyist versions. Rotation of the work is driven by the headstock, which is clamped to the work; the tailstock acts as both a bearing and a support for the opposite end of the workpiece.

The classic spring-return drive system for a wood lathe retains this alternating power and recovery stroke. But it frees both of the woodworker's hands to manipulate cutting tools and any other relevant devices. This spring-return lathe came into use sometime before the year 1000 CE, and it remained a normal item in a wood shop until perhaps 1900—coexisting with various cranking arrangements and external power sources. Museums of old technology commonly house and demonstrate examples.[9] As in figure 7.8, it adds two components to earlier designs—an overhead spring and a bar below the lathe that the woodworker rhythmically steps on.

My biomechanical bias gives me a peculiar interest in the spring returns. Two different kinds have coexisted. One amounts to nothing more than an archer's bow, resized as convenient now that it

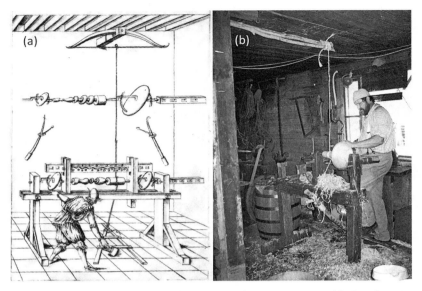

Figure 7.8. (a) Jacques Besson's bow-spring-driven ornamental lathe of 1578. Note the complex shapes being followed. (b) A bent-sapling foot-powered lathe, in my photograph taken at a demonstration at the Minnesota State Fair of 2007.

needn't be limited by the roughly 30-inch maximum draw distance of an archer, as in figure 7.8a. The other is a bent pole, from which a string or strap descends to wrap around the turning shaft, as in figure 7.8b. Most accounts describe the pole as a "sapling," implying both an unsliced and a fresh bit of tree, by contrast with the carefully prepared wood of the bowyer. Now, this detail doesn't just contrast crude with sophisticated technologies. Wood as tree differs greatly from wood as prepared material. Think about how you tell if a small branch is alive or dead. You do a simple mechanical test, bending it—if alive, it bends readily, if dead, it snaps. From a very early age, we who climbed trees and kindled campfires were familiar with the test. Using a bow-type drive bends a piece of wood far less than one might bend a sapling. But it takes advantage of the greater stiffness of dead wood and, of course, the long-term stability of prepared wood.

Using a sapling bends the wood more, offsetting the lack of stiffness while taking advantage of the greater tensile strength of fresh wood. So the bow stores energy as more force and less distance; the sapling as less force and more distance.

To finish the business of lathes, I ought to note that the reversing rotation of spring returns suggests that the cutting tool be cyclically drawn back inasmuch as it contributes only friction during coun-terrotation. As a practical matter, this made reversing rotation im-practical for metal lathes, where very rigidly supported cutting tools had to be advanced into the work smoothly and steadily. Cranks, waterpower, then whims, and steam engines made metal lathes into the practical machines that, without much exaggeration, can be said to have permitted the industrial revolution.[10]

Far less common than return strokes driven by springs are ones driven by energy held as gravitational potential, that is, as raised weights. At first glance, gravity seems ever so handy. It's available ev-erywhere with a degree of uniformity matched by few other things in nature. Taking advantage of gravity asks only that a weight be raised during the powered stroke to store up what's needed for the return. One can raise a small weight to a great height or a heavy weight to a modest height with equal effectiveness, and simple pulleys permit free choice within that spectrum of possibilities. The device can be quite small. One of our closet doors persisted in drifting open when drawn by an air current from our front door. A string, a few hooks, and a descending weight well hidden in the recesses of the closet provide a fix that has yet to be noticed by a guest.[11] Or the device can be huge—a trebuchet hurling stones, dead horses, or plague-ridden corpses as a five- or ten-ton weight descends.

A number of things incorporate pulleys that rotate first in one direction and then in the other, incorporating just this form of grav-itational storage. For instance, houses with conventional windows, the kind that you opened by lifting the bottom half, once used coun-terweights to ease the task of lifting wood and glass in an all too friction-afflicted track. But rotation of their pulleys itself plays no central functional role. Rather, it's just a way to change the direction of a force exerted along the length of a rope or cable—a solid beam may handle forces in various directions, but ropes and cables resist only lengthwise tension and gravity pulls insistently downward. Put another way, simple pulleys exist because you can't push on a rope.

Figure 7.9 shows one of those windows, with a rope (or chain) extending upward on each side of both lower and upper panes; it passed over an inset pulley as it disappeared into the wall. Within the wall, long cylindrical pieces of cast iron descended as each window went upward—so-called sash weights of almost 10 pounds. Those pulleys were none too good to begin with and, between accumulation of dirt and repeated misapplications of paint, they eventually became simply frozen tracks for the cords. The cords responded by

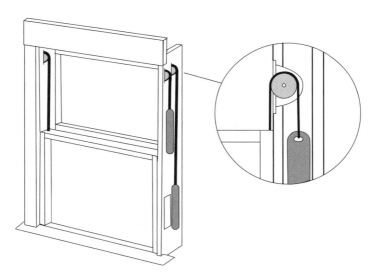

Figure 7.9. Sash weights and pulleys on a conventional double-hung window, that is, one in which both panels can be raised and lowered. The weights turn up in junk shops and can serve as door-closing weights, as mentioned, often more satisfactory than springs. Never mind recent impostors—this is the classic windows operating system.

fraying and, all too frequently, breaking. The latter event usually happened as a window was raised or lowered and was announced by thunks and clunks as that hidden sash weight plunged down until it hit some lower bit of house framing. In old houses, the repair demanded fairly invasive carpentry; newer construction provided mildly helpful access panels.

Perhaps our current reluctance to design windows with counterweights has something to do with the contemporary practice of building as little as possible at the site of end use, an inevitable side effect of the economic attractiveness of mass production. A counterweight simply must be heavy; a spring can store far more energy relative to its own mass. That weight economy of a spring presumably more than offsets the counterweight's nicely steady force, not a force that varies with extension, bending, or twisting, as in a spring. And, especially for intermittent tasks, we care little for energy economy. So my garage door opener has a motor large enough to raise it unaided, wasting the energy made available as the door descends. A pair of pulleys and counterweights (perhaps in vertical tubes) would permit use of a smaller motor, and, of course, a minor bit of energy economizing. In fairness, though, one has to admit a drawback of that constant force for many potential applications. A descending weight at some point hits bottom. Whatever kinetic energy it retains—and if descending it must have some—will ordinarily be wasted in the impact, whether sudden or buffered. That's one reason why counterweight trebuchets could never reach the efficiency of sinew-sprung ballistae.

If we don't insist that the task itself be rotational, then examples of counterrotation driven by gravitationally stored energy can be recognized in pre-industrial human technology. Figure 7.10 shows a scheme for drawing water up from a well or river with steep banks. A donkey or mule or pair of bullocks pulls on a rope that extends over a pulley and thence down to bucket and water supply. Instead of pulling horizontally, the power source walks down a prepared slope, gaining help for the task from its own weight. Freed of the

Figure 7.10. Raising water from a well with a traction hoist—simple and effective, given a draft animal, an effective yoke, and a reasonable pulley. Arranging for the animal to walk down slope when loaded has long been a common way to increase effectiveness (from Ewbank, 1842).

weight of water, it turns, walks up the slope, and lowers the bucket again, putting energy (again from its own weight) into storage for the next pull. What goes around with no net rotation is that passive item of necessity, the pulley. Several other traditional water-raising techniques make analogous use of counterweights and thus stroke-to-stroke gravitational energy storage, typically with only a small arc of rotation of beams in either direction beyond the repeated rotation and counterrotation of pulleys.

To persuade myself just how ordinary were these arrangements, I improvised a muscle-powered bench grinder (for metals) that operated in much the same manner as a spring-pole lathe, but with a counterweight instead of a spring pole—a bench grinder as might have been made in past centuries (fig. 7.11). Construction presented no great difficulty, but I then recognized how spoiled I've been by the high revolution rate of a motorized grinder, something not practical with one that repeatedly reverses its rotation. The figure shows its construction—a bearing block, a long bolt for a shaft, a grindstone, a rope, and a weight. Operation consists of pulling down periodically,

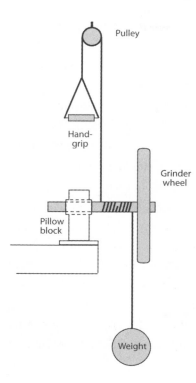

Figure 7.11. My counterweight-return hand-operated grinder. The shaft diameters of the pull rope and weight rope attachments can be different if one wants to play force-distance trade-off and mismatch pulling force and weight.

either on the stroke during which the object to be ground is pressed against the stone or on the stroke making the stone turn in the other direction. The choice depends on how the rope is wound and thus is the grinder's option.

Of the types of energy storage with which we power reverse rotation, only inertial storage remains, saved for last as the least obvious and thus the one that we're most surprised to find both ancient in origin and widespread in use. The principle here is that of the flywheel—something that, once set into circular motion, persists in going around. So the work put in to make it spin can be drawn out again at some later time. Imparting angular momentum, the relevant variable, differs in no important way from imparting linear momentum, what a weapon imparts to a projectile. For linear momentum, the amount of momentum you've imparted to the projectile equals its mass times its speed. For angular momentum, the amount of momentum reflects the speed of rotation and how far the mass lies from the axis of rotation. As noted back in chapter 3, the relevant formula contains not just that second part, the distance from the axis of rotation, but the square of that distance—mass gives a lot more angular momentum if located well away from the center of the spin. That's why flywheels on machinery typically take the form not of uniform circular disks, but of wheels with thick peripheries.

Where might you notice flywheels in action? Automobiles have them deep inside the engines, working to smooth out the necessarily pulsating pushes of the individual cylinders. Old steam engines, particularly those used on farms, had only one or two cylinders; they depended on flywheel smoothing and mounted their big flywheels out in full view, as in figure 7.12. Small toy cars (Hot Wheels and so forth, vehicles with so-called friction engines) that keep rolling well after an initial push, do so with tiny flywheels, the latter hidden

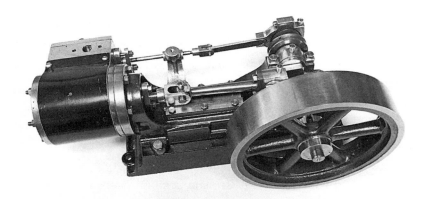

Figure 7.12. A model steam engine, this one a single-cylinder, double-action design, as might have been used (if full-size) for farm machinery. "Double action" refers to a cylinder pushed by steam in both directions, equivalent to a four-cylinder Otto cycle automobile engine—and yet it still needs a flywheel. This is one of many built by a neighbor, Ed Dougherty.

beneath the outer casing and set into rotation by the initial motion of the cars' wheels.

A moment's thought may (or may not) bring to mind a major stumbling block for making inertial storage reverse some rotational motion. You spin a flywheel, and it keeps going in the direction in which you've spun it. That's clearly not the direction you need to reverse what you've done. The trick, then, consists of making that angular momentum wind up the string, cord, or strap that has just been unwound—with no reversal. This positions it so the next toss or stroke causes the device to turn in the other direction. Put another way, a string initially wound in a right-hand helix rewinds itself in a left-hand helix (think screw threads and pushing on a Yankee screwdriver). So the next operation reverses the earlier rotation.

All these words—what about an example? The most familiar, with no doubt whatsoever, must be the yo-yo (or properly but uncommonly "Yo-Yo," since the name was registered as a trademark by the Duncan company). A yo-yo (fig. 7.13a) consists of a pair of circular disks connected by an axle, with a string looped around the axle. The string is wound up around the axle, and the player holds a loop on the free end. Dropping or, much better, throwing the yo-yo unwinds the string, spinning the paired disks. The disks then continue to spin, and the string (after, if need be, a tiny jerk to start it winding) winds itself back up, now wound in the opposite direction, as just explained. Thus the next toss spins the thing in the opposite direction. That's typically obscured by the yo-yo's left-right symmetry as held in the hand or by the way the player regrasps the toy after it makes an inconspicuous 180-degree rotation at a right angle to its axis.

So familiar are yo-yos that one fails to notice a critical facet of their operation. If you were to throw a stone tied to the end of a string, the point at which the string was pulled taut would announce its arrival with a sharp jerk. By contrast, when you throw a yo-yo, it gradually slows down and very gently reaches the full extension of its string. What has happened really defines yo-yoing in physical terms. You put some linear momentum into the system with your

Figure 7.13. (a) The modern yo-yo, basic version, which gives every indication of being the same as the (b) 480 BCE version from ancient Greece.

toss. Momentum may be conserved—that's a basic physical law—but it can be transformed. Here it's transformed from linear into angular form as the yo-yo rotates ever faster, and so the device slows down after its initial gravitational or operator-induced acceleration. Incidentally, gravity needn't play much of a role. Throwing a yo-yo downward may be the easiest toss, but an adept player can throw it almost as easily in any direction.

The yo-yo has remarkably ancient roots. Most sources suggest a Chinese origin, but with clear examples from illustrations on pottery and surviving terra-cotta disks (fig. 7.13b) of apparent diffusion to classical Greece by 500 BCE. They were well known in both India and the Philippine Islands a few hundred years ago. The present incarnation traces its ancestry (and name) to the latter, through a small company in California, the Yo-Yo Manufacturing Company, opened by a Filipino immigrant in 1928, and bought out soon afterward by Duncan Toys. In the past fifty years, designs and contests have proliferated, and technological advances (and complications) have moved yo-yoing far beyond the pastime of youngsters.[12]

Yo-yos, however high-tech, have always been and still remain toys—notwithstanding claims that they may have been used by arboreal hunters. But the principle they so nicely illustrate also drives a thoroughly utilitarian device of still greater antiquity and still wider distribution. This paragon of ingenuity usually goes by the name "pump drill," although some accounts call it a "push drill," confusingly the same as a drilling version of the Yankee screwdriver that's still commercially available. Like a bow drill, a pump drill serves equally well—and about equally as often—as a frictional fire starter and as a mechanical drill.

For most of us, appreciating the way the device works requires reference to an illustration (fig. 7.14); even better is a look at one of the many videos posted on the web.[13] The subtlety, though, should be minimized by all the talk about the yo-yo, its partner in inertial momentum and energy storage. Four components must be present: (1) a shaft, either with a bit on the end for drilling or a friction pad for starting a fire—in no particular way different from the shaft of a bow drill; (2) a crosspiece with a hole that fits loosely over the shaft, one long enough so each end can be grasped in one hand; (3) a cord or strap that runs up from one end of the crosspiece to an attachment at the top of the shaft and down again to the other end of the crosspiece; and, finally, (4) a flywheel of wood, stone, or pottery located below the lowest possible position of the crosspiece. A few variations occur. Some Chinese pump drills put the flywheel on top, reducing stability but presumably giving a better view of the work. Some pump drills substitute a transverse beam with stones at each end for the flywheel—simpler to make, perhaps, but requiring deliberate dynamic balancing before use. Small pump drills need only one hand on the crosspiece for operation, with the fingers straddling the shaft.

Operation begins by turning the crosspiece around and around until the cord is fully wound and the crosspiece has been raised nearly to the top of the shaft. The driller then pushes down on the crosspiece with both hands as symmetrically as possible, which turns the shaft. The cord unwinds and then, like the string of a yo-yo, winds

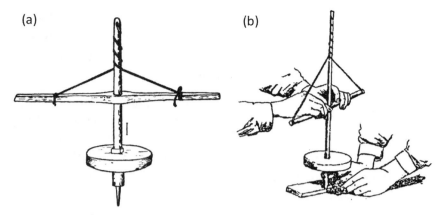

Figure 7.14. Two pump drills, where (a) has the crosspiece extended sufficiently far that the hands operate outboard (from McGuire, 1896), and
(b) positions the hands inboard of the attachments of the cords (from Hough,
1890).

back up on the shaft with the opposite direction of twist. The driller
then repeats the stroke, and the shaft again turns, although this time
it goes in the opposite direction.

In a way, a pump drill has only half the effectiveness of a bow
drill because useful pressure on the tip of the shaft will be exerted
almost entirely when the crossbar is pushed downward. If starting a
fire, then, the pump drill works half-time; if drilling with a unidirectional bit, it works only quarter time. Not that quarter time should
be sneezed at—the pistons of our automobiles, with their Otto cycle
engines, give a push only during a quarter of their total travel, during
the downward half of the power stroke.

No simple analogy with anything like archery could have inspired
the creation of the pump drill. Its antiquity has never been in question, but both its specific time and place of origin remain obscure.
While evidence of the use of pump drills in ancient Egypt may be
ambiguous, worldwide distribution among pre-industrial societies
(at least in recent centuries) has been thoroughly documented. One
has the impression that no regional ethnological museum would

consider its collection comprehensive without an example. Trouble comes, though, in judging which examples represent acquisitions subsequent to contact with modern European and Oriental explorers, traders, and settlers. Accepting those ambiguities, then, pump drills have been reported from Inuit, Aleut, and Haida in the North American North and Northwest, from various Native American tribes elsewhere, and from Madagascar, Siberia, Indonesia, Polynesia, Melanesia, Micronesia, New Zealand, and Australia.[14]

Back in 1896, J. D. McGuire described a simple and effective drill he had made and tested that combined features of bow drill and pump drill.[15] It required the top support of the bow drill, but used the revolving flywheel to unwind and then rewind a string, as in figure 7.15. He called it a "top drill" after toy tops, and he drew connections with some figures drawn by the ancient Egyptians. I don't find the connections all that persuasive, but, conversely, I think the model must have found some use; whether or not ancestral or derived, something so simple must have occurred to someone, somewhere.

But, after building a model (about which more in the appendix), its relative uncommonness surprises me less. A pump drill avoids applying the top drill's awkward sideways force, so centering demands less instant-to-instant adjustment in the force applied by the hand

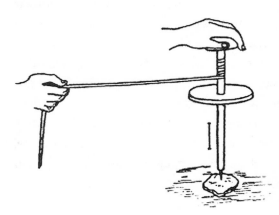

Figure 7.15. J. D. McGuire's top drill—simple, effective, and unknown among any of the diverse human societies.

at the top. And the pump drill automatically relieves the downward force on the drill so the thing can wind itself back up, while the top drill asks the operator to deliberately alter the downward force from half cycle to half cycle. In short, a top drill works, but it demands more skill and attention than does a pump drill—which, in turn, asks more than does a bow drill.

Two final notes: First, we've seen four schemes to supply energy for counterrotating—muscle, elasticity, gravity, and inertia. These four ways to end up with no net rotation of a rotational device don't by any means exhaust the range of possibilities. At least in theory, any way to store energy will suffice. For instance, power companies sometimes build what they call "pumped storage" installations to smooth out the load imposed by consumer demand from their generators—during periods of low demand, they pump water into reservoirs located on some convenient mountain; when demand increases, release of water turns turbines that generate additional power. Nothing prevents the trick from being applied to some muscle-powered device— nothing except practicality.

And second, an item that didn't fit especially well at any specific point earlier in this account. We sometimes find it handy to design a rotational machine with no overall rotation, one in which rotation of one part is balanced by counterrotation of another part. All too often, unbalanced rotation can be a nuisance—or worse. The familiar example of dual, opposite rotation is the eggbeater, whether it's hand operated, a portable electric, or a stand mixer. (Single-beater stand mixers do occur, but they turn fairly slowly and have steep-sided bowls.) I've seen an electric knife sharpener with a pair of counterrotating abrasive wheels. More generally, counterrotating pairs of cylinders commonly occur in devices for squeezing and crushing. The everyday example a generation or two ago was a clothes mangle, the pair of rollers mounted atop a washing machine that squeezed water out of fabric before proper drying.

[8]

The True Crank

Finally—that most familiar, most obvious, most ubiquitous, and seemingly simplest way to convert the circular translation that a muscular appendage does so well into true and continuous rotary motion. I speak of the crank, as found on hand-operated kitchen gear such as eggbeaters and on foot-operated vehicles from bicycles to human-powered aircraft. More specifically, we'll mainly talk about human-powered cranks, as in figure 8.1, ones in which some handle or pedal turning parallel to an axle is connected by an arm to that axle, so that moving the handle or pedal around the axle, even if just translating it around, will turn (crank) the axle.

Cranks, familiar enough, have lost what preeminence they might have had in our particular household. Since we two adults no longer own bicycles and since small electric motors have colonized every one of our rooms, the number of true cranks in our household has dropped to seven—unless one stretches the etymology and includes a pair of curmudgeonly aging humans.[1] Still, a room near my university office houses a few basic metalworking tools such as a lathe and a milling machine and contains over a dozen quite obvious cranks.

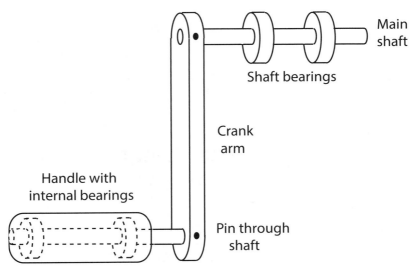

Figure 8.1. The basic elements of a hand- or foot-operated crank. Translation of the handle in a circle produces true rotation of the shaft.

Speaking of the etymology led me to wonder whether the origin of the term "crank" as applied to this wonderful contrivance differed from "crank" and "cranky," pejoratives for forms of human unpleasantness. A link between these latter two and the German *Krankheit*, for "sickness" was obvious and easily confirmed. It turns out that the mechanical term comes from the same root, one that in Old German and Old Dutch meant "bent," among its other connotations. Modern German avoids the ambiguity by using *Kurbel* for the mechanical device, nothing that will be mixed up with *Krankheit*.

Bow drill and pump drill go back to the prehistoric times; wheel-and-axle devices may be more recent but came early, at least in the Old World. Cranks, seemingly as simple, have a sharply contrasting history. One might imagine that someone would have envisioned the advantage of adding a lateral extension to the periphery of the wheel on a wheel-and-axle, and a technological takeoff point would have been reached soon thereafter. But for some reason the trick took hold relatively late and after what appear to be some nearly abortive starts. Whether the classical Greeks and Romans knew nothing of cranks or merely used them only on rare occasions seems to me a minor distinction. In any case, I have no strong opinion; to use the crude local phrase: I don't have a dog in that fight. The question just becomes whether, in classical antiquity, they were negligible or non-existent. J. G. Landels, extolling Greek and Roman engineering, disparages the utility of cranks relative to handspikes.[2] I don't fully buy his argument, especially in the light of how cranks largely supplanted handspikes in cultures that knew both and cannot be credibly accused of impracticality.

Tomb pictures from ancient Egypt show what looks suspiciously like a crank on a drill, as shown in figure 8.2, one reproduced in many accounts. Weights on either side of a vertical shaft press it

Figure 8.2. A bas-relief of what provides the best claim that the ancient Egyptians drilled with true cranks, from Saqqara, about 2450 BCE.

downward, and the right hand of the operator looks poised to either stabilize or turn the shaft. The operator's left hand holds what might be a crank handle at the top, a curved secondary piece extending outward and then upward to one side of the axis of the shaft. V. Gordon Child makes the case that the thing doesn't work as a crank at all, and that opinion seems to be the general consensus.[3] We may presume that the operator is right-handed, so that hand does the main job, to repeatedly shove one of the weights (push and release, again) around the shaft to turn the drill. What does the left hand do? Child suggests that it merely grasps a supporting piece, perhaps a cow horn, something crucial to keeping the shaft vertical. I'd add that if the curved top piece were truly a crank handle as we know it, the eminently practical Egyptians would have fashioned it in a shape more convenient for that function, with more of a vertical section. And they would have added some contrivance to keep the whole thing from wobbling uncontrollably, something a hand on the side stones wouldn't have accomplished.

Nonetheless, the Egyptian device may have functioned as some kind of crank, but the story leads into some waters deeper than it seems best to plumb at this point; we'll return to it later in the chapter.

The earliest definitive evidence of cranks as we know them comes from northern China, in the second century BCE. Turning a crank turned a fan that winnowed grain, here wheat and millet, and later (in the south) rice—winnowing separates the grain from the chaff by blowing away the latter. A fan eliminated the need for tossing the mix in the air on a windy day, and a crank-operated fan could be (and reportedly was) a portable and thus rentable tool. The evidence is good—models, some with working parts, from ancient tombs. Cranks then spread to a wide range of devices such as the windlasses of wells and querns.[4] In the West we have an illustration (repeatedly copied) from 850 CE of a pair of individuals operating a very large cranked grindstone; one turns and the other holds a blade against its peripheral surface.[5] That seems to be the earliest definitive image in Europe of a crank as we know the device.

By the time of the great mechanical compendia of the fifteenth and sixteenth centuries, the ones I keep mining for examples, cranking technology had spread far and wide. More to the present point, it was now doing all kinds of tasks in all kinds of permutations. Mariano Taccola, in *De Ingeneis*, illustrates two cranks, put in diagrammatic form in figure 8.3.[6] The more conventional one (to our eyes, at least),

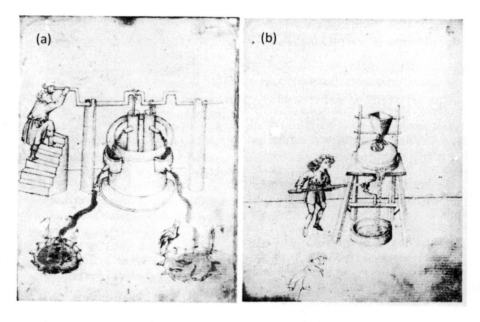

Figure 8.3. (a) An early illustration, by Mariano Taccola, of a conventional crank and crankshaft running a pair of piston pumps. Note the obvious error in the illustration—the pump on the right would not pump because the connecting rod is attached on the axis of the shaft and thus would experience no reciprocating motion. The connecting rod should instead be attached to the unoccupied crank on the right. (b) A crank turned, not directly, but by a pushrod to operate some form of mill.

figure 8.3a, runs a piston pump; the crank attaches outboard of a two-bearing crankshaft, with a rod mounted on an offset segment of the shaft going up and down (as well as translating around) with each turn of the crank. We'd now call that rod, the thing that converts rotary to reciprocating motion, a "pitman arm"; it will reappear later in the chapter. The other (fig. 8.3b) reverses this action. The operator doesn't crank at all but rather pushes and pulls on a rod whose other end then translates around the offset segment, the crank, of a vertical shaft. The shaft, in turn, runs a mill of some sort. The action is that of a piston in your automobile as it turns the crankshaft that eventually drives the wheels by pulling up and pushing down.

That pushrod of Taccola (fig. 8.3, again) occurs occasionally in the other compendia, but it often comes with a significant addition. Attached to the end of the rod is another, pivoted at one end and free at the other, so that, with the pushrod as load, it forms a second-class, force-amplifying lever, as in figure 8.4. The operator can then push and pull a greater distance than the swing of the crankshaft but needn't exert as much force as ends up driving the machine.

The illustrations provided by several of Taccola's successors give a wonderful window into the richness of their cranky world. The figures in Agricola's book, one in which mechanical devices are only a part, include several dozen cranks. Ramelli shows no fewer than sixty-six. Besides direct, human-powered cranks, some of their machinery has lots of internal cranking linkages, places where we might use timing belts or even trains of spur (= ordinary) gears. Each of these designers, as well as Strada and Besson, occasionally put a crank on both ends of a shaft; usually that demands two crankers, although Ramelli illustrates several portable devices in which one person turns both cranks, one with each hand—these particular ones were intended to tear fortified doors from their hinges.[7] In all dual-crank devices, the cranks operate exactly out of phase, that is, the arm of one points upward when the other points downward. The designers clearly recognized that this arrangement not only doubled the power (for two crankers) but smoothed the action—just the way

Figure 8.4. One of the many figures in Strada's compendium showing a crank worked by a second-class lever; here it turns the lantern of a crown-and-lantern gear pair. One wonders whether its presumably wooden elements could handle the high forces or whether key elements—pushrod and crankshaft—had to be metallic. Also, note the flywheel on the lantern gear shaft (below the crown gear) to keep the shaft turning at the extremes of the crank's travel (where the line of action of the lever is unfavorable).

we hook up the pistons and crankshaft of a two-cylinder engine so they run in exact opposite phase.

A common alternative to a second crank is a set of handspikes on the end of a shaft opposite the crank, as in figure 8.5. What might initially seem an odd mechanical miscegenation makes good sense for these serious power-transmission systems. For the most part, we're looking back to a world of really energetic cranking, not, as

Figure 8.5. Combining cranks, for rapid, low-force motion, with handspikes, for more forceful but slower movement: (a) as done by Agricola, for raising and lowering a bucket at a mine shaft, and (b) as done by Besson, for operating a large vertical press.

at present, for control devices (on lathes, for instance), for intermittent operation (rechargeables), or for toys. A crank serves well as a way for a human to power a machine if the crank can be turned fairly rapidly against a load that doesn't vary much with the crank's rotational position. Cranks are at their best if the load has lots of angular momentum relative to its frictional resistance, as has a flywheel, for instance. Bicycles should be pedaled at between 60 and 90 revolutions per minute for greatest efficiency, hence our fixation on gear choice. For lower rates of revolving and for irregular loading, handspikes come out ahead. Where the loading regime varies from time to time, what could be better than provision of both ways to couple human to machine?

And that's consistent with where this odd pairing occurs. For instance (fig. 8.5a), for one of the ways that Agricola proposes to raise ore from a mine, handspikes should be especially handy for lifting up a large filled bucket; the crank just as certainly should be more than adequate to lower it when emptied.[8] Besson (fig. 8.5b) shows the two on a huge triple-screw vertical press, an application where, for greater efficiency, we'd now use a hydraulic press.[9] The crank would do nicely for lowering the pressure plate fairly rapidly until whatever was being pressed started to fight back; then the second person working the handspikes could take on most of the now more slowly proceeding task.

One peculiar practice mars one's admiration for the practicality of these designers. The crank arm does the simple task of connecting the handle with the parallel but not coaxial driveshaft. The job takes nothing more than a straight bar with appropriate end fittings. But, as noted in the translator's discussion at the end of Ramelli's compendium, a belief persisted among many (but not all) designers that a curved crank arm worked more effectively. In particular, a good arm had its C-shaped curve extending rearward relative to the direction of rotation, as in the ones in figure 8.5b and figure 8.6. The oldest cranks, from 850 CE, as well as those of Taccola, had straight arms. Agricola used straight arms for hand-operated cranks but, oddly, severely curved arms for cranks (operating as pitman arms) within mechanical linkages. By contrast, Ramelli almost always employed curved crank arms. Zonca, likewise, used curved arms, but Strada kept his straight.

This design detail remained contentious for at least two hundred years, but curved crank arms persisted long after that. I noticed that an early nineteenth-century hand-operated book press in the library of my home institution had a curved crank arm. Lest we dismiss that as premodern, I offer as example the twentieth-century hand grinder in our kitchen, a yet newer device with at least a modestly curved crank arm. One might, perhaps, make the slightly forced argument that the curve of the arm of an otherwise severely utilitarian design at least guides the operator's choice of the direction in which to turn the handle. What few cranks incorporate is a tapered arm, broader near the central axle, where the forces are greater, a design for which a good functional argument can be made.

This era of proliferation of cranked machinery in Europe, roughly what we call the medieval period, saw the development of small as well as large cranks. Our familiar (or up until recently familiar) brace and bit originated then and (except for the bit holder) has changed very little since. Not all applications could be classed as constructional, military, or industrial. A curiously sophisticated, hand-cranked musical instrument appeared in the twelfth century, the hurdy-gurdy (figure 8.6). This should not be confused (as it often is) with the barrel organ of the iconic organ grinder and monkey—the latter (the organ, not the monkey) a later invention, also hand cranked but merely to run an air pump. In a hurdy-gurdy, turning the crank turns a wooden friction wheel whose outer surface, coated with rosin, rubs against a taut string just as does the bow of a violin, except that it demands no pause to change direction of bow stroke. A set of finger-operated levers, a keyboard, adjusts the tension on the string and thus determines the note. Faster turning increases volume, leaving frequency unchanged, again as in the present stringed instruments that we play with bows.[10]

The prize for the most audacious application of human hand-cranking must go to the submarine *Hunley*, of the navy of the Confederate States of America, during the American Civil War. The

Figure 8.6. A fairly bare-bones hurdy-gurdy with a double-curved arm on its crank. Notice the friction wheel on the same shaft as the crank. Courtesy of the Duke University Musical Instrument Collections, Gillian Suss, Curator.

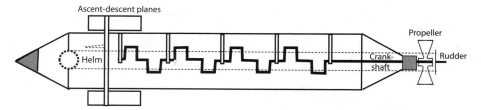

Figure 8.7. The drive system of the CSS *Hunley*, in top-sectional view. Control lines are shown dashed. See also www.hunley.org and an excellent *Wikipedia* article.

Hunley has the distinction of being the first submarine to sink an opposing ship, in its case the USS *Housatonic*, in the harbor of Charleston, South Carolina, in 1864, although the *Hunley*, with eight crewmen, was itself sunk in the action. In the ship (see fig. 8.7), 39.5 feet long but a mere 4.25 feet high, a long crank shaft had eight cranks (although only seven seem to have been manned; the eighth crew member took care of the controls and armament). Never mind the claustrophobia—the thing was a death trap, losing all but three of its crew in three separate sinkings.[11] Not that the *Hunley* was the first hand-cranked submarine to mount an attack. During the American Revolution, David Bushnell's *Turtle*, a one-person craft with a hand-cranked vertical auger (as well as cranked propelling screw) tried and failed to drill a hole in the hull of a British warship in New York harbor.

Nothing other than human hands, aided by arms and torsos, urges any one of those fourteenth- and fifteenth-century cranks to turn. Yet humans are primarily walkers and runners and only incidentally climbers and minimally competent as brachiators, so we have a far larger fraction of our body musculature devoted to moving hind limbs than forelimbs. Treadmills and related devices do good work and, as we've already seen, have served us long and well. Can human hind limbs provide power through any other coupling system,

← Crank arm

← Treadle

Figure 8.8. A classic spinning wheel. Often a cord substitutes for the crank arm leading upward from the treadle and isn't preserved. (Smithsonian.)

perhaps one that's not so large and cumbersome as a treadmill or treadwheel? The obvious thing would be a foot-operated crank. And the simplest form of foot-operated crank might be a treadle, as on an old-fashioned sewing machine. The long, pivoted rod that helped turn a crank in figure 8.3b, a feature capitalized on by several of the designers, comes close, but the crank remains hand-operated. Besson's lathes have foot pedals, but these either lift weights or stretch bows, with distinct recovery phases.

So when did the crank-turning treadle, whose essential elements are shown in figure 8.8, make its appearance? Treadles made an early appearance in one particular application, driving spinning wheels, and they did so not all that long after the introduction of the wheels themselves. Spinning wheels appeared in China, India, and the Islamic world by the thirteenth century and in Europe a little later, but these earlier versions appear from surviving illustrations to have been hand-turned. Treadles, whatever their specific date of invention (1533 is sometimes given), were common by the later part of the sixteenth century. Beyond spinning wheels, terminological ambiguity complicates any attempt to trace the spread of the applications of treadles, in particular the way the term "treadle" came into use for any kind of foot pedal. For instance, so-called treadle looms were introduced to Mayan weavers by the Spanish after the conquest. But these treadles lifted and lowered warp threads rather than turning cranks. In any case, by the eighteenth century, treadle-driven cranks drove a diversity of devices. Small wood lathes could be turned by the feet of the same person who applied the cutting tools. Any problems of history and terminology should not obscure the great advantage of the crank-turning foot-operated treadle—a single operator then retained free use of both hands.

Another curious musical instrument also took advantage of the way a treadle drive freed both hands—the glass harmonica, which first appeared in London, in 1761. The instrument hasn't quite faded into oblivion, both because of the general fame of its inventor, Benjamin Franklin, and of the excellence of several composers who provided scores—Handel, Mozart, and Beethoven in particular. Franklin's glass harmonica (or armonica) had a size-graded set of bowls mounted axially on a horizontal shaft, all turned by a wheel, which was in turn turned by the crank and treadle (fig. 8.9). The player touched the rims of the turning bowls with wetted fingers, just as

Figure 8.9. A glass harmonica in an illustration made shortly after Franklin's practical version appeared in 1761.

one might run a wetted finger around the rim of a partially filled wine glass to generate a fairly pure tone. Changing the rotation rate changed the intensity of the sound.[12]

The last major machines that had treadle-driven cranks are, of course, sewing machines. They still do good work where electric power isn't available. I recall my mother's old Singer as it operated in the late 1940s—a previously used machine that she had bought ten years earlier and whose most recent patent dates were around 1915. It differed little from its electrically driven replacement, the latter purchased after an attempt to electrify the old one overstressed its internal mechanisms. Isaac Merrit Singer[13] invented this kind of treadle-operated sewing machine, the first one suitable for household use, in 1851, although his patent was held inferior to that of Elias Howe's, of 1845. The treadle-crank itself was by then too well established to be considered on its own.

What must be our most common and efficient way to crank has yet to put in an appearance—foot-operated pedals. When we want to couple human to power-demanding task with the greatest possible efficiency, in particular in contriving human-powered aircraft, we've consistently turned to foot-pedaled cranking. But this terrific scheme began, so to speak, only yesterday, more specifically, in the latter half of the nineteenth century. That is, if we ignore an almost certainly bogus claim that a student of Da Vinci produced a startlingly modern-looking (but wooden) bicycle in 1493.[14]

To me, the late appearance of cranks, at least in Europe, appears surprising; conversely, the very late appearance of foot-cranking seems much less so. It happens to be efficient, but it's neither mechanically obvious nor biomechanically natural. Put bluntly (and literally), turning two cranks, one with each leg, with the cranks operating out of phase with each other leaves you without a leg to stand on. You can't even mobilize your back for each cycle as you can when rowing or paddling—incidentally also efficient couplings of person to power-demanding task. Feet and legs don't normally translate in circles in any of our various gaits—walking, running, race-walking, hopping, or skipping. Neither walking nor running wins prizes for mechanical efficiency. They do derive some help from, respectively, gravitational and elastic stride-to-stride energy storage, so the repeated need to decelerate and then reaccelerate limbs isn't quite as bad as it might be. By contrast, pedaling takes little or no advantage of such energy storage, either by lifting the body (gravitational) or stretching tendons (elastic).

Only the long familiarity of our chief pedaled device, the bicycle, lets us forget just how strange and unlikely are its various bases—in addition to pedaling. Not only is pedaling peculiar, but so is its system of steering, with a necessary offset between front axle and fork and the combination of handlebar turning and rider tilting, not to mention the oddity of needing only two wheels to stay upright without acrobatic agility on the part of the rider. No surprise, then, that the modern bicycle evolved in several steps, mainly in the second half of the nineteenth century. One can't imagine anyone conjuring up such a machine *de novo*.

The first step established the practicality of a pair of wheels, fore and aft—the way it resisted tipping over as long as it remained in motion. Going by various names, these velocipedes or hobby horses (fig. 8.10a) provided a seat for a rider, who propelled the unit by alternate (as usually depicted) or perhaps synchronized pushes of legs against ground.[15] At least on hard surfaces, even the heavy contraptions of the early nineteenth century permitted higher speeds with less effort than did any pedestrian gait. One can imagine the unexpected pleasure of coasting downhill, inescapable proof of the stability of the two-wheel design. About going uphill, one finds little mention.

When the evolving bicycles moved beyond ground pushing, pedals didn't at first drive the wheels. The earliest linkages did crank, but treadles drove the cranks. Exactly who was first to build a functional no-foot-on-ground bicycle remains murky; the main contenders are two Scotsmen, confusingly MacMillan (in 1839) and McCall (in 1869). Their treadles differed from those we met earlier as drivers

Figure 8.10. (a) A velocipede or hobbyhorse, this one a steerable model.
(b) A treadle-bicycle; the fellow with the beard is McCall himself, in this
1869 illustration. (c) A penny-farthing; its awkwardness needs no comment.
(d) The modern bicycle, this one a Columbia of 1896. (All from the Smithsonian Bicycle Collection.)

of lathes and sewing machines—a pair was operated out of phase,
with one leg on each, like modern bicycle pedals. The arrangement
depended on the fact that the rider's body was perched on the bicycle's seat (fig. 8.10b). Other than the treadles, the McCall machine
(about which we know more) looks fairly modern, with two wheels
of equal size, the front one steered and the rear one driven (one can't
say "pedaled," of course). Treadling with feet operating antiphase
bears an obvious resemblance to our normal gaits; one wonders how
it felt in practice.

Pedals, meaning circular translation of feet in the now-familiar
fashion, appeared in the 1860s, in both patents and commercial

products. One has to remind oneself just how novel and, in a sense, unnatural were foot-driven cranks. Applied to bicycles, pedals initially drove front wheels, the obvious choice for a direct connection of pedals to axle. That led to a gradual shift from wheels of equal size to what in the end became the so-called penny-farthing bicycles (after the relative sizes of the British coins) with front wheels as large as human legs could possibly straddle and rear ones only as large as needed to traverse minor bumps, as in figure 8.10c. Speeds seem to have increased, riders rode higher and needed more skill, and an accident caused greater injury.

By the end of the century, the modern bicycle (fig. 8.10d) appeared and rapidly displaced all earlier designs. The vehicle was composed of two wheels of equal size with aerostatic tires and a set of pedals amidships that drove the rear wheels by way of a chain. Its earliest name, the "safety bicycle," pointed up the well-recognized hazardousness of the earlier form. It hasn't fundamentally changed since, although present models offer countless augmentations, improvements, and items of fashion—better brakes, tires, and general ergonomics, more versatile gearing, superior materials, and so forth. The chain drive offered high efficiency from the very start, and it has come in for relatively less alteration than other components.

Exactly why the system of foot-operated cranking makes such good use of human effort remains a little hard to pin down. It does bring to bear a large mass of muscle, most of which is normally active in the consummately aerobic activity of distance running—meaning the right kind of muscle. But that can't be the entire explanation, as a few numbers will illustrate. A good power-producing muscle can give an output of about 200 watts per kilogram for a short time and perhaps sustain an output of 100 watts per kilogram.[16] Muscle makes up about 40 percent of the body mass of a lean human. If a 70-kilogram person works a quarter of that muscle, 10 percent of body mass, maximally (and, as bipedal runners, our muscle is concentrated in the lower body), then an output of 700 watts should be sustainable. But we can't keep that up for more than about half a minute. Sustainable output, say in the range of 10 minutes to an hour or so, of even highly trained, gifted athletes does not exceed 300 watts. So where's the rub? The limit ultimately depends on what the cardiovascular system can do, mainly as it supplies the muscles with oxygen. The implication of that 700 versus 300 watts is that the specific choice of muscular action and mass of muscle brought to bear may matter, but only at the margins. What makes more difference is the somewhat unexpected excellence of the coupling of those large masses of muscle to the task. Proper gearing allows the rider to operate at rates that maximize power output. Racing bicycles and human-powered aircraft must represent the highest efficiency level of this coupling.[17]

Finally, just how good are bicycles? If we take as our measure the metabolic cost to go a given distance, then cycling is far cheaper than any pedestrian gait. At 4.5 miles per hour (admittedly a rather fast walk), cycling is 2.2 times less costly than walking. At 9 miles per hour (6.7-minute miles), it's 3.7 time less costly than running. All this despite an increase of 10 to 20 percent in the weight you're hauling around, since even the best bicycles don't come close to weightlessness. Still, a comment made earlier in connection with wheeled vehicles should be kept in mind—such data presuppose hard-surfaced roads. In terms of cost per distance, the worse the path, the better walking becomes relative to cycling. Mountain biking may be fun, but it's only for the aerobically able and cardiovascularly confident.

Several promissory notes from earlier in the chapter ought now be redeemed—I put them off so their discussions wouldn't divert us too far from the main story.

First, that purported ancient Egyptian sort-of crank. The main source here is a rarely cited but fascinating paper by A. W. Sleeswyk.[18] He points out that two bearings (or sets of bearings) support typical cranks, thus keeping their shafts rotating around fixed axes. One uncommon kind of crank—exemplified by the lasso, hurling sling, and bola—has no bearings and takes considerable skill to manipulate. In between, and almost never recognized as such, are cranks that have one bearing, so-called partially guided cranks. The nearest thing to a familiar example might be a pair of children playing jump rope, with one end of the rope tied to a tree and a child cranking the other end. Sleeswyk reinterprets the tomb illustration shown earlier as figure 8.2 as a partially guided crank, and he offers a model and a proper physical analysis of its performance.

As an inveterate model-builder, I could not resist building a version of Sleeswyk's model—functionally equivalent, if physically handier for the materials and tools I had at hand; figure 8.11 shows the two versions. The thing seemed implausible, and the first few minutes of my initial trial definitely bore out that impression. But I quickly found that it worked surprisingly well—at least for shallow holes. Wobble diminished as skill increased, the whirling weight

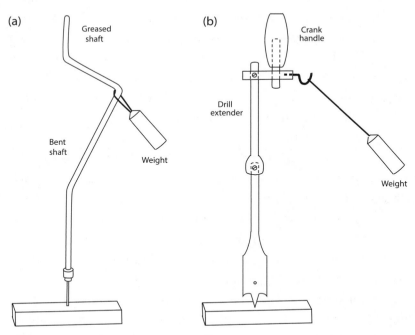

Figure 8.11. Two versions of a functional interpretation of that ancient Egyptian drill, Sleeswyk's (a) and mine (b).

meant that while pushing downward may have been impossible, it really wasn't necessary. The circle around which my hand had to translate was surprisingly small; and, unlike a brace and bit, the device left a hand free to steady or adjust the work.

I look at both Sleeswyk's drawing and my model and then at the tomb illustration and still remain less than fully persuaded that we're really imitating the actual Egyptian drill. The drill may be remarkably effective and the Egyptians remarkably clever, but the resemblance doesn't come close enough to assert with confidence a proper relationship. The tomb illustration has two weights; the second would just make trouble. The weights in the illustration don't seem free to swing about any wide arc. The handle at the top turns in a considerably wider arc than I found necessary. And no left hand should be needed to turn the weights. I hope I'm wrong and Sleeswyk's right — I've become inordinately fond of my model.

And then a few other mechanical bits and pieces of cranking systems. The pitman arm received brief mention earlier, and one recognizes its opposite, modestly known simply as the connecting rod. After all, sometimes we want to connect a rotary prime mover, say an electric motor, to a reciprocating task, perhaps a sabre saw or sliding-blade electric razor. At other times we want to connect a non-rotating prime mover such as a reciprocating piston or even a muscle to a rotary task.

A piston going up and down or back and forth in the cylinder of an engine, as mentioned when we took note of Taccola's cranks, differs in no important respect from a muscle contracting repeatedly or a person alternately pushing and pulling on some handle. It makes no difference to the analogy that the piston in your car actively pushes during less than a quarter of its travel while that of a double-acting steam engine may push almost continuously. Muscle, for that matter, operates in a mode in between, actively shortening and then passively being lengthened, with the latter job done by some external agency such as another muscle. Muscle happens to be a short-stroke engine, shortening by only about 10 percent or less of

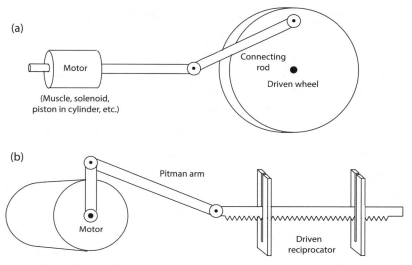

Figure 8.12. (a) A reciprocating motor drives a rotational device, as in a steam engine or automobile. (b) A rotary engine drives a reciprocating device, as when a water wheel drives a cross-cut saw.

its resting length when called on for reasonably high power output—closer to what's done in the engine of your car than to the action of a classical steam engine. But, while close, these analogies aren't particularly enlightening; the only thing really worth comment is that all these engines require connecting rods to convert reciprocating to rotating motion, as in figure 8.12a. In cars, the connecting rods are sometimes called piston rods; in steam engines, the two are distinct components.

The opposite type of component may be of less direct relevance to the present discussion, but it carries a better story and in any case completes the account. If you have a waterwheel and want to drive a saw back and forth or, as was more common in past centuries, up and down, you need to convert rotational motion into reciprocating motion. Now the component, visually indistinguishable from the connecting rod, is called a "pitman arm" (fig. 8.12b). Notice that the name begins with no capital letter (some *Wikipedia* entries and many other sources notwithstanding). Yes, people do carry the surname "Pitman," but none of these has eponymic status.[19] How the pitman arm functions was what the term originally described—it directly replaced the pit man, the person down in the pit beneath the heavy piece of timber being sawed into planks by two sawyers, one above and one below.[20] (A counterweight or similar device later frees the upper sawyer.) This particular linkage has been traced back to the Roman Empire,[21] so Taccola was showing nothing startlingly new.

Four final notes to this whole story of cranks: First, the early steam engines, such as the mine-pumping engines of Newcomen, not only themselves reciprocated, but they drove reciprocating devices. So nothing turned continuously. Applying steam to rotational machinery was mainly the achievement of James Watt, late in the eighteenth century. Initially his engines did not have the connecting rod linking to crankshaft of the ordinary and by then obvious kind. Watt worried about a preexisting patent on that arrangement and so instead used a peculiar kind of epicyclic gearing arrangement.[22] Retrospectively, it seems most peculiar that something that at that point was about 500 years old should have merited patent protection.

Second, going back just a few years, piston engines of the kind introduced by Watt weren't the first thermal engines that converted their linear motion into rotational motion. In one application, the earliest practical steam engine, Newcomen's rocker-arm pumping machine, was coaxed into doing just that at the suggestion of the greatest of early civil engineers, John Smeaton, of Eddystone Lighthouse fame. But the arrangement, perhaps making a virtue of necessity, now strikes us as distinctly odd. In the huge Carron Ironworks, in Scotland, around 1760, air was blown into furnaces with large pistons. These, in turn, were run by waterwheels, presumably with the pitman arms just described, which depended on running water. A Newcomen pump, by recycling the water, eliminated that riverine prerequisite; by turning the waterwheel (albeit indirectly), it converted its linear into rotational motion. So linear motion became rotational, which became linear once again. In addition, some of the earliest electric motors reciprocated, in particular the magnetic rocker of Joseph Henry, in 1831, sometimes considered the first proper electric motor. (More about this motor in chapter 11.) Thus for most tasks, it would have required conversion, although by its time that would have been routine. In any case, rotational motors quickly eclipsed any promise that Henry's motor might have had.

And third (really, reiterating), cranks can serve as coarse, rapid, forceful adjusters to be followed by grasp-and-release fine, lower-power adjustment, as when a lathe tool is first brought to and then applied to the work. Or they can serve as lower-power, rapid adjusters to be followed by more powerful handspikes, as when a press is lowered to touch and then to put the squeeze on something. In either type of application, the cranks' virtues are the speed and steadiness of movement they permit.

Finally, a note on the direction of cranking—and the turning of cranked devices. Most of us have a dominant right hand and arm. We find overarm cranking easier, and we can develop more force doing that rather than underarm cranking. That's not just from long experience—to quote a proper study of the issue, "reverse arm cranking requires greater muscle activity from the biceps brachii, deltoid, and infraspinatus muscles."[23] Early automobiles required that the operator crank the engine as part of the starting ritual. For a crank attached (temporarily) to the front of a front-to-back longitudinal engine, overarm cranking by a right-handed person makes the engine turn clockwise, when viewed from the front. But, unlike the standardization of some other things that reflect their early history, engine rotation has not consistently retained that clockwise rotation even in the remaining vehicles (such as most pickup trucks) that have longitudinal rather than crosswise driveshafts within their engines. At least that's what I found in a very brief survey—in the process amusing as well as puzzling some acquaintances who wondered why I cared at all.

[9]

Spinning Fibers

We turn now to the process with which, had this book taken a strictly chronological approach, we would most likely have begun. Long before wheeled vehicles or potter's wheels or rotary querns, probably even before bow drills, humans began to spin short fibers into long threads and ropes. The name does not mislead—"spinning" implies rotation. It can be done with unaided hands, but even very simple appliances, devices in which no parts move one against another, can greatly facilitate the task. Thus, in that admittedly limited sense and even if lacking any proper wheel and axle, these earliest spinning aids must represent the first rotational machinery.

While spinning has generated a wonderfully rich literature— archaeology, textile history, industrial history, fashion, and handicrafts—I find one curious dog-that-didn't-bark. Almost none of the sources that I've looked into pay attention to the underlying question of what in a mechanical sense spinning accomplishes. What does this peculiar self-enwrapping action ultimately yield? Perhaps much can be put down to perceived irrelevance, but where mechanics does come in for mention, I suspect misapprehension. As put by M. L. Ryder, "The crucial point is that one cannot weave a strand comprising a mere aggregation of fibres because such a strand lacks strength owing to an absence of cohesion."[1] But no basis for cohesion is immediately evident, and the surface irregularities often alluded to as consequential don't seem necessary—one spins smooth as well as rough fibers with much the same mechanical outcome. The real story, at least as I understand it, is too good, too revealing, and of too much present relevance to leave out.

A laughably simple demonstration, shown in the photographs of figure 9.1, gives a hint of what's going on. Roll a cloth napkin or handkerchief into a loose cylinder and run water over it until it's quite wet. Then pull on the ends of the cylinder—most likely almost nothing will happen, which is why I include no specific figure. Then, again loosely, twist the rolled cloth lengthwise by a few revolutions. Now, as in the figure, pull once on this helical version—it will compress laterally and most likely will drip water along its entire length. Think about what the twisting has done to the individual strands that make up the cloth. If it has a normal weave, that is lengthwise and crosswise, twisting makes the strands run helically, with one set stretched out a bit and the other going just a bit slack.[2] Pulling on the cloth pulls on the extended set of helical strands; they respond by moving toward the center of the cylinder; they try to take a more direct path from end to end. The slack set just gets slacker—once again, you can't do much by pushing on a rope. So the whole thing shrinks, and it tells you that it's shrinking by squeezing out water.

Now the most basic problem in making serviceable cordage—from threads to cables—out of natural materials is all too common and all too basic. Precious few such materials come in lengths sufficient to serve directly for any of our ordinary uses for cordage. Leather strapping,[3] silk, a few sorts of stripped underbark, dried and treated macroalgae—a few, but just a few. Wonderful fibers abound in nature, but they're short, whether the cellulosic ones of the plant kingdom such as flax, jute, cotton, and the like, or the proteinaceous wools and other such hairy animal stuff. How can long ropes be made out of short fibers? The answer, discovered empirically, consists of

(a)

(b)

Figure 9.1. Pulling on a wet, slightly twisted cloth napkin with a normal crosswise weave. In (a) the napkin has been thoroughly wetted by dipping a loop in a bowl of water; in (b) I'm pulling on a rope tied to one end of the napkin; the other is held in a bench vise. Water appears as glistening and dripping, a result of the lateral compression produced by pulling on the helical winding of one set of threads in the napkin.

spinning the fibers, running many of them in parallel, but twisting those parallel arrays into helices. Then the harder you pull on the resulting long helix—call it a thread—the harder those fibers squeeze against each other just as did that cloth handkerchief or napkin. And the harder they squeeze against each other, the more strongly they resist pulling apart. Never mind the famous Indian rope trick of magicians—this is the real rope trick.

We might call what we're looking at "dynamic gluing." Friction, which hasn't been mentioned in the past few chapters, now plays a critical if unheralded role. While silk and our synthetic fibers may be smooth, many of our most useful natural fibers are anything but smooth—they're irregularly kinked, like cotton, or rough surfaced, like wool. So they don't like to slide ("shear" in mechanical jargon) along one another, even less so if pressed together. Spinning, that is, twisting the fibers together, engages this wonderfully automatic mechanical action—the harder you pull lengthwise, the harder the fibers press against each other crosswise. But even smooth fibers, which mean most of our synthetics, will show the effect since they still develop friction as they pull along one another. Pull on a piece of knitting yarn or rough twine and the sideways compressing effect should be obvious. Subtle and wonderful—shear stress between adjacent fibers.

Another way to look at the situation provides a nice touch of irony. For one old technology, making wheels turn on axles, a key part of the game consisted of reducing the friction caused by one surface shearing against another—recall chapter 3. That's what lubrication accomplished. For this other old technology, spinning long cordage from short fibers, key to success lies in assuring high-enough friction so that shearing forces are sufficient to keep the fibers from sliding apart. If resistance to sliding motion alone, the critical role of friction, leaves you skeptical, try breaking a length of cotton thread fresh off the spool and then breaking another whose center has been lubricated with a drop of oil.[4]

For spinning, rotation thus becomes an absolute requirement, which brings us back once again to our basic biomechanical conundrum, making non-rotational muscle drive rotational machinery. For spinning, a solution exists that's both particular and universal, one that's either trivially obvious or marvelously subtle depending on how your intuition happens to operate. Better than just a set of drawings is the simplest of demonstrations, one that takes only a sharpened pencil, a foot or so of string, and an inch of adhesive tape. Tape one end of the string to the middle of the pencil, and then loosely roll most of the rest of the string around the pencil by twisting the pencil. Now ease the string off the sharpened end of the pencil without further rotation. You will have made a tangled mess, the result of introducing a number of full twists equal to the number of times you turned the pencil. I once bought a cheap reel for kite strings that worked this way; it gave no trouble at all the first time

I flew a kite; but, since the reel could not be reversed on its axle, it became next to useless thereafter.

If in place of that string you start with a loose bundle of rough fibers of, say, cotton or wool, then the rolling on and pulling off in this particular way twists them into a strand. The result yields quite an impressive increase in tensile strength. Perhaps I ought to offer a more specific demonstration, as in the sequence in figure 9.2. As it's packaged, absorbent cotton, the kind that comes wrapped up in rolls and is sold in some drugstores and medical supply stores, has fairly well aligned fibers, eliminating the preliminary step called "carding," which you may have to do if you use batting from a fabric store or hobby shop.[5]

Assuming commercial absorbent cotton, strip three lengths of at least 20 inches from the roll, with each piece about three-fourths of an inch wide. Instead of a pencil, use some cylinder roughly half an inch in diameter, part of a duster, coat hanger, or other household item — only one end need be free. Tape one end of a cotton strip to the cylinder, a few inches from the free end, and loosely wind the rest on by turning the cylinder. Then, holding the other end, ease the coil off the end; as with the string described earlier, it will have acquired a number of twists equal to the number of times you rotated the cylinder. Wind it on and ease it off again to double the number of turns, and then tape the twisted strip down on some convenient surface. If it's not taped down, of course, it will spontaneously untwist — not surprisingly the fibers, being proper solid materials, are elastic, which means they will be eager and at least partly able to reverse any applied stress.

Now do exactly the same thing with the other two strips, so you have three strips side by side, each with the same direction of twist. Examine the twist. If it goes in the direction of the threads on an ordinary screw, the direction in which a clockwise turn would make the screw go into some object, we call the coil a right-handed helix, or Z-twist — the middle segment of the letter Z gives the twist's

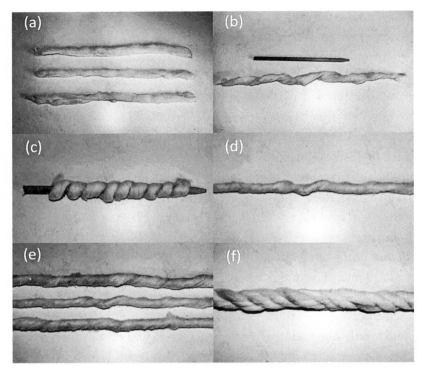

Figure 9.2. Making a cotton rope. (a) Three lengths of combed absorbent cotton. (b) A twisted strand after one roll onto the stick and a pull off its end. (c) The same, rolled onto the stick again. (d) And the strand, now pulled off again and thus further twisted. (e) Three such strands. (f) Finally, the three, wound together in the opposite direction to form a stable and surprisingly strong cord.

direction. If it goes in the opposite direction, the one in which a clockwise turn would loosen a screw (or jar lid), we have a left-hand helix, or a so-called S-twist, for the direction of the middle of the S (or from "sinister," meaning "left"). Then tape one end of each of the three nearly identical strips to the cylinder and wind the rest of the three on together. Here's the critical part. This second winding has to be in the opposite direction from the first. Slip the doubly coiled three-strand cord off the end of the cylinder and secure its ends with loops of tape, independent of any other surface. Giving coil and supercoil opposite twists greatly reduces the tendency of the resulting structure to spontaneously unwind or throw itself into awkward tangles.

You've made a cord—a structure that has far greater tensile strength than the three strips of short, loose fibers with which you started—by doing nothing more than reorienting those fibers. That, in short, is the great rope trick. Sometimes a single winding of a few fibers suffices; sometimes yet another stage of coiling may be added, but the game remains the same.

The origin of spinning remains—and will most likely remain— obscure. In part, one can blame the usual reluctance of the key materials to persist for our *post facto* examinations. Nonetheless, extant scraps of material from a wide variety of places attest to the early development of spun fabric; they clearly point to an origin more than 10,000 years ago. The other complicating factor is the uncertain relationship between the spinning of fibers and four different ways spun fibers might be employed—making nets, baskets, ropes, or fabrics. The first two, netting and basketry, demonstrate in unambiguous fashion the way the in-and-out, over-and-under of weaving flexible material produces something of great utilitarian value, whether as fish traps, sieves, or carrying containers. But baskets, at least, and probably the simplest of nets, don't require that their material be spun. For that matter, we still make baskets of unspun plant material that has simply been soaked to render it flexible enough to weave.

Ropes and fabrics can certainly be made of long vines or strips of tanned hide without recourse to spun fibers, and fabrics can be fashioned as felts (perhaps with the help of some binder). But we have long preferred to make them with nicely twisted stuff—by contrast with basketry. Basket- and net-making necessarily demonstrate both the technique and advantage of weaving, but neither necessarily provides a requisite step in the evolution of spinning fibers for subsequent weaving.[6]

As the little demonstration with pencil and absorbent cotton showed, spinning a long, tension-withstanding thread from short fibers could not be easier. Strictly speaking, even the pointed stick isn't necessary—hand against hand or hand against thigh will induce a twist, something a pastry baker does once in a while with dough. No, the practical problem with spinning has a somewhat different basis. To expose what the task is that is asked of the spinster (in Middle English, the word literally meant "woman who spins"), I made a few measurements and some quick calculations. A bed sheet of no particular pretension (from Ikea) has thread with about one full twist (360 degrees) per millimeter and 80 threads per inch in each direction. A queen-size sheet measures 90 by 102 inches. Sparing you the details, that means a set consisting of top sheet, fitted sheet, and two pillow cases has absorbed roughly a million full turns. Spinning with some analog of that pencil, doing two turns per second (faster, perhaps, while turning but pausing to shift spools, orient new fiber and so forth), would take 17 eight-hour shifts. The end user would have to pay $1,360 just to provide a minimal wage of, say, $10 per hour to the spinster, never mind any other cost. The problem, then, comes from the vast amount of spinning absorbed in making any significant amount of fabric.[7]

In both Eurasia and the Americas, as far as we know independently, a much faster way to spin long ago supplanted hand against thigh, which survives only within a few isolated human populations. While many variants exist, the basic arrangement can be fairly simply described and illustrated (fig. 9.3). A flat, toroidal whorl is attached near either the top or bottom of a long, cylindrical spindle. The spindle is almost always wooden; the whorl may be made of wood, but some other denser material such as stone works better. An unspun length of gathered fiber supports the dangling spindle and whorl, so the appliance can't be too heavy. The spinster holds in one hand a batch of prepared but as yet unspun fiber; that batch may be loosely rolled on another stick, called the "distaff"—another word we've appropriated for contemporary redefinition. The other hand

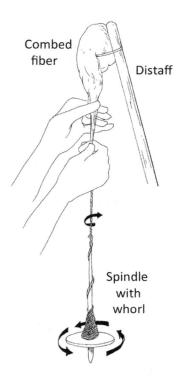

Combed fiber

Distaff

Spindle with whorl

Figure 9.3. A generic version of the basic arrangement for hand spinning. Spinning the spindle and whorl with one hand twists the loose fibers between the spindle and the other hand; the thread produced is wound on the spindle and the process can continue indefinitely. (Drawing from Tiedemann and Jakes, 2006, © Margaret Davidson 2015; www .margaretdavidson.com; reused with permission.)

is free to spin the spindle and whorl. Spinning the latter twists the unspun length of fiber, which is then wound onto the spindle. A new unspun length is then extended and the process repeated.[8]

The persistence of stone spindle whorls in the archaeological record prompts a parenthetical note. At least among us non-professionals, our view of prehistoric antiquity must surely be biased by that persistence of stone, somewhat greater than metal and far greater than wood, bone, horn, or natural fiber. Wood, in particular, lends itself to shaping more readily than does stone, as well as being lighter (less dense), tougher (less prone to cracking), and springier (more resilient). What we call the "stone age" could be a tacit misnomer, with our youngsters even more misled by Fred, Wilma, and the rest of the Flintstone family than we ordinarily suppose.

The whorl, of course, acts as a flywheel, something we've met before, but here burdened with a peculiar physical constraint. Wider is better for providing the angular momentum that keeps it going, but too wide and it will interfere with the operation, especially if located near the top of the spindle. Angular momentum depends on weight, but too heavy and the unspun, extended length will break. The best whorl would be one that had its weight concentrated peripherally, that is, one with a thin middle and thick rim. But either from lack of understanding or from the inconvenience of fashioning such a whorl, such an optimized design turns out to be rare. Whorls, incidentally, provide persistent artifacts for archaeologists. For instance, the presence of iron spindle whorls in Southeast Asia in the early centuries BCE indicates the presence of Indian cotton weavers—as noted by Judith Cameron of the Australian National University, "These are hard to use without knowledge and have no intrinsic value."[9]

As the quote implies, to avoid breaking the unspun length and then to make thread of decently uniform thickness demand considerable skill, something acquired through long practice. Spinning finer thread takes a thinner spindle as well as more turns per unit length of thread. The fineness of cotton fabrics spun in India was legendary in Europe, where coarse spinning of woolens was the general practice.

While a hanging hand spindle turned out thread a lot faster than did thigh spinning, it can't be described as fast relative to the job at hand—it still required a disproportionate investment in human (almost always female) labor. Sometime before the year 1000, most likely (once again!) in China, a device appeared that increased the rate at which spindles turned by perhaps a factor of ten. As in figure 9.4, a large hand-turned wheel drove the much smaller spindle by way of a tight cord. Concomitantly, the spindle axis shifted from vertical to horizontal. This combination of a so-called great wheel of large diameter and a spindle of small diameter represented no radical mechanical novelty, but it ushered in a technological revolution

Figure 9.4. A hand-turned spinning wheel, this one as drawn about 1480 (reproduced in Patterson, 1954). The flywheel is located on the right; the spindle and bobbin are located on the left (where spinning actually occurs). In operation, the left hand would be used to feed fiber into the bobbin while the right spins the flywheel.

that in turn has been assigned credit for major social changes. Further improvement (getting a little ahead of the story) came with the addition of the foot treadle in the sixteenth century (as described in the last chapter), freeing the second hand of the spinster. And then came further elaborations—flywheels, bobbins, and so forth—that permitted continuous spinning with no requisite pause to wind the thread onto some other spool.[10]

By the thirteenth century, spinning wheels of the basic kind were in wide use throughout Eurasia. All still asked that one hand turn the wheel and that spinning periodically pause to wind the newly formed thread off the spindle and on to some kind of bobbin. The great historian of medieval technology, Lynn White, considered the consequences in Europe.[11] Relative to other commercial commodities, the price of cloth, that of linen in particular, dropped significantly. All that high-fashion starched and folded linen of late medieval and later years became practical. More significantly, linen rags—and clothing inevitably wears out—made excellent paper, so manuscripts could be written on paper rather than on the always costly parchment.[12] If paper had not been available, Gutenberg would have had no economic incentive to develop his printing press, nor, if he had made one anyway, would it have rapidly proliferated.[13] Book availability and literacy—a correlation doesn't seem forced. Literacy and social change—again the correlation doesn't stretch credulity, and White viewed wheel-spinning as one of the medieval inventions that enabled both and therefore enabled the European Renaissance. Book burning has long appealed to thought-controllers; it lacks effectiveness only because books are too numerous, too durable, and too easily concealed.

Even more. All those sails catching wind for the boats plying the Nile—those of the classical Mediterranean cultures, and the longboats of the Vikings, even those of the great Polynesian boats—represented a vast investment of the labor of simple hand spinning. Each of these kinds of boat, whatever their diverse features, carried one sail of modest size and relied much of the time on oarsmen or paddlers. Beginning shortly after spinning wheels made their way to Europe, larger boats were developed, ones entirely dependent on sail, ones carrying far greater areas of cloth on multiple masts— and much more cargo. Exposed to wind, weather, and continuous handling, sails wear out all too quickly. One can make a reasonably persuasive case that the great era of sail and discovery depended to a considerable extent on the lowered price and greater availability of sailcloth, and that in turn came from the revolutionary way that spinning wheels made spindles turn faster.

Anthropologists have gathered enough data to provide a good quantitative and comparative view of the problem facing pre-industrial textile producers. For making a Peruvian poncho, from collecting fiber to completing the basic textile, spinning represents nearly 70 percent of the total investment of time—a number obtained when a simple spindle does the job. And spinning is neither the slowest nor the fastest way to produce thread by hand. Starting with wool (and comparing different and only loosely comparable studies), thigh spinning produces thread at 15 inches per minute, spindle spinning achieves about 40 inches per minute, while spinning with a treadle-driven wheel does almost 175 inches per minute. In effect, wheel spinning is four times faster than spindle spinning and twelve times faster than thigh spinning.[14] Put yet another way, even wheel spinning can't quite reach 3 inches per second and would take nearly four hours just to make the thread for our set of bedding. For comparison, individual units of large modern spinning arrays go at more than 100,000 revolutions per minute, producing thread twenty times faster still.

We've made woven textiles by spinning a large number of natural fibers; the particulars have been determined as much as anything by local availability. South Asia long ago domesticated cotton; another species of *Gossypium* was later domesticated, spun, and woven in South America. Egypt and the Middle East developed flax cultivation for linen production. Europe early on collected the spring shedding of underfur and then sheared sheep and some other animals, while llamas and their relatives provided analogous wool in the New World. Within all this diversity, only two basic materials played major roles, as noted earlier—the insoluble carbohydrate cellulose in the plant-derived fibers and the protein keratin (as in fingernails and bovid horns) in fur and wool. One mineral fiber, asbestos, saw limited use; ancient writers marveled at a fully fire-impervious fabric.[15]

What prompts this mention of the specific starting fibers is an odd exception. One natural fiber, silk, doesn't require spinning at all, since the insect larva extrudes a single glued pair of filaments, about a thousand meters long, to form its cocoon. But the silk of the ordinary silkworm, *Bombyx mori*, comes in strands only 10 micrometers in diameter (a hundredth of a millimeter), which is awkwardly fine for most of our purposes. So for a very long time we've loosely twisted several of these paired filaments together to bring the resulting thread up to a manageable thickness. Despite circumventing any critical role for spinning, silk production ends up being even more labor intensive than making fabric from other common fiber sources. Silkworms have to be reared on mulberry leaves. As with most leaves, these aren't exactly nutrient rich, so an individual larval insect has to eat a huge mass of leaf before producing its cocoon. Then it takes about 2,500 cocoons, cooked to degum the fiber and then unwound by hand, to produce a pound of silk. One source suggests a labor investment of almost 40 hours for that pound, which makes spun fiber look cheap, whatever the particular spinning technique.[16] Silk production, however uneconomical, has ancient roots, beginning in China at least 3,000 years ago. For obvious reasons, it has always been a luxury item.[17]

Introducing silk, the non-spun textile, permits a digression that's at once entomological and etymological. We speak of a silkworm "spinning" its cocoon, a complete misnomer inasmuch as nothing spins—the process consists entirely of extrusion. Why the peculiar mislabeling? Almost certainly because spinning was the process by which we made thread, and the silkworm made thread. It gets crazier yet. When William Kirby and William Spence published their classic entomological text, between 1815 and 1826, they called the extrusion organ of the silk moth the "spinneret," because it made the silk filament from which we made the thread, never mind that it didn't spin.[18] Still worse. When artificial fibers were first produced, in the late nineteenth century, the extrusion nozzles were also called "spinnerets" from the obvious analogy with those of the silkworms. These industrial spinnerets didn't spin either; artificial fabric may have been made from spun thread, but spinning happened at another stage in the process, meaning it wasn't done by the spinnerets.

None of the devices introduced so far can make more than thread, yarn, or thin cord. Rope of any substantial diameter simply can't be spun hand against thigh, with a dangling spindle, or with a treadle-operated spinning wheel. So, while rope-making may present the same basic physical problem, it doesn't share the same technological history. Nevertheless, the antiquity of ropes matches that of woven fabrics, which amounts to declaring both prehistoric. The ancients depended on ropes for a multitude of tasks, and only a few kinds of cordage don't require the twisting trick that makes long tensile elements out of short fibers. Strips of leather cut as spirals from a hide yield cords many feet long. A few vines can stay usably flexible after cutting, some particularly flexible (if perishable) in underwater applications. The Indians of the Pacific Northwest of North America made some cordage by treating the stipes of local macroalgae, which can be tens of yards in length.[19] But these are the exceptions.

Even pre-dynastic Egypt made good twisted ropes, and the construction of all the monumental structures of both Old and New Worlds depended on ample supplies. Not that all the ropes had to be spectacularly strong. From Egyptian carvings it's clear that, in dragging heavy loads, often each worker pulled on a separate rope. Since draft animals didn't provide much if any assistance, nor did the Egyptians make use of pulleys, that meant tensile forces of about 100 pounds (roughly 500 newtons). The numerous yards that kept the masts of their sailing ships erect must have sustained greater forces, but their large numbers should have meant that individual ropes didn't sustain extremes. Ancient Egyptian vessels plying the Nile put particularly severe loads on the great cables that ran from stem to stern, the "hogging hawsers" (or hogging trusses) that, as in figure 9.5, kept the boats from bending upward (hogging) amidships.[20] But each boat needed only one or at most a few of them.

The quantities of cordage absorbed by old technologies are mind-boggling—if one thinks about it at all. Of course, the supreme examples of large-scale projects remain the Egyptian pyramids, for which rope consumption must have been, one is tempted to say, monumental. Still, no single artifact exceeded the great sailing ships

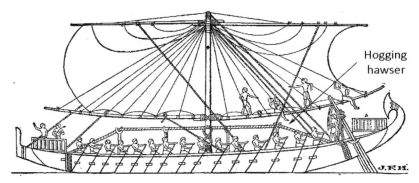

Figure 9.5. An ancient Egyptian boat, with its hogging hawser, as shown (with a note pointing out the feature) in H. G. Wells's *Outline of History* of 1920—something I read long ago.

in the lengths of rope required at once. Horatio Nelson's flagship, the HMS *Victory* (launched 1765), had a complement of about 20 miles of rope. A first-rate ship of war in 1866 had 43 miles of rope, weighing almost 80 tons, and during the War of 1812 (and Napoleonic War), the Royal Navy had over 800 ships—admittedly most of them of considerably lower displacement. Nor were the ropes manipulated only for routine maintenance—sails had to be raised, lowered, furled, trimmed, and so forth. One has no trouble envisioning the likely origin of the phrase "learning the ropes."

And ropes could be large; if we're to believe the Greek historian Herodotus (of admittedly irregular reliability), when the Persian king Xerxes invaded Greece, his army crossed the almost mile-wide Hellespont on two floating bridges held in position by a dozen ropes of flax and papyrus, each 9 inches in diameter. Isambard Kingdom Brunel's largest steamship, the *Great Eastern* (launched 1858), had a rope 15 inches in diameter—it was made, incidentally, of coir, familiar to us moderns as the material of natural doormats.

An exhibit at the Chatham Historic Dockyard, southeast of London, gives some sense of the scale of these old naval natural-fiber

Figure 9.6. The undone end of a manila rope. According to the label, 1,944 unit threads combine to form nine primary strands, which, in turn make three cable strands.

ropes as well as reminding us of the process—figure 9.6 is a photograph I took a few years ago of one of their exhibits.

Making anything from a thread to a cable requires that the item be kept under tension while it's being formed. For making thread or yarn, either the weight of the spindle or the pull of the spinster's arm will be adequate. For ropes, providing that tension takes (in the literal sense) considerably more effort. The practical consequence has been that, until recently, ropes were made in extended form, stretched out as they were being made. In the simplest arrangement, fibers were first spun into yarn by one of the spinning techniques already described. Further twisting then took place in a long, straight ropewalk, one of which is shown in figure 9.7. Lengths of yarn, typically six or more, were payed out and twisted together, often by a person turning a stick while walking backward and leaning away from the forming strand to give it the necessary tension. Two or more of these strands were then twisted together in the same manner, but with the opposite direction of twist to minimize the rope's incentive to undo the process

Figure 9.7. The ropewalk of the Plymouth Cordage Company, Plymouth, Massachusetts, early in the twentieth century.

when released.[21] We might prefer to do the twisting with something other than a stick turned hand over hand, that is, with grasp-and-release cycles. Perhaps we might hold a stator board in one hand through which a crankshaft extended while turning the crank with the other hand. But, as we've seen, the ancients weren't big on cranks.

Ropewalks, then, were standard industrial items wherever ropes were made in any quantity. During most of the past millennium, that meant where ships were built. In consistently warm and relatively dry climates, any flat outdoor strip of bare land would do, and these left no direct traces of their existence. By contrast, in northern Europe and North America, rope-making took place in dedicated buildings, in particular in specially built long, skinny buildings. Enclosure became ever more important as mechanization increased. In 1810 the United States, then little more than a settled strip along the East Coast, had 173 ropewalks. Henry Wadsworth Longfellow (1807–1882), a New Englander living during the heyday of wooden shipbuilding in that region, found the local structure worth a poem (1858), which begins:

> In that building, long and low,
> With its windows all a-row,

Like the port-holes of a hulk,
Human spiders spin and spin,
Backward down their threads so thin
Dropping, each a hempen bulk.[22]

In the eighteenth and nineteenth centuries, no organization had more sailing ships than the British Royal Navy. So, unsurprisingly, they built a particularly impressive ropewalk, one that still stands and can be visited at their Chatham Historic Dockyard, a short train ride southeast of London. When constructed in 1728, it was the longest brick building in Europe, long enough to lay out a thousand-foot length of rope.[23]

While talking about ropes, a few words about their practical peculiarities might be added. First, and the point should be obvious, ropes sustain only tensile forces—you can't usefully push on a rope. So, unlike beams, columns, and shells, their cross-sectional shape doesn't influence their strength; what you see directly reflects the way a particular rope was made.[24] But that same inevitably tensile loading imposes a peculiar risk. If you put a compressive load—a squeeze—on something, a crack in it may not make too much difference. The load itself will tend to keep the crack closed, so failure will be unlikely. By contrast, loading in tension—stretching—opens any crack that makes its appearance. That reduces the area of material bearing the load, which may let the crack propagate farther across the material. Thus, once started, a crack is all too likely to lead to catastrophic failure. Besides the obvious advantage gained by its flexibility, making a rope do duty as a tension-resisting structure provides some insurance against this kind of failure. A crack across one fiber or one piece of yarn or even, with multi-strand rope, one strand doesn't find it at all easy to jump across to the next fiber or strand and continue on its evil path.

This insurance against easy crack propagation, as well as the assurance of flexibility, underlies our practice of using twisted multifilament ropes. We do so even where the filaments—as with silk, steel, or the various extruded polymers—are themselves very long so the twist no longer results from some necessary spinning. It also explains an odd hazard of depending on ropes that are exposed to cold, wet weather—as were those of all too many sailing ships. Any ship that ices up not only suffers from all that weight high above its center of buoyancy, but its rigging can be weakened by the way the ice facilitates crack propagation across individual ropes. Fortunately, ice happens to be a lousy solid material, itself cracking easily, so it usually (but not always!) fractures first. An analogous but less serious hazard afflicts washed clothing hung out to dry at subfreezing temperatures.

Finally, back to that basic business of enlisting shear—interfilamentous friction—in making long, strong rope from shorter fibers. Even with that trick, longer fibers have an edge over shorter fibers. Manila fiber, from a wild banana plant (abaca) native to the Philippines, makes especially good natural rope, in part because of the great length of its individual fibers, from 4 to 16 feet. Or perhaps one should say "made" good rope—natural rope having become largely anachronistic in a world of synthetics in which component fibers can be of any length and the result resists the usual agents of rot. In the end, though, no matter what aspersions we cast at ropes made from natural materials, I can't fail to notice that I'm wearing cotton clothing—shirt, shorts, underwear, and socks—all products for which a spinning process lies at the start of their manufacture.

[10]

A Few More Turns

A few more aspects of circular motion ought to come in for attention, as much as anything to illustrate its multidimensionality and versatility. These might have taken pride of place in an early chapter, but that would have ultimately diverted us from the main line of the present story, essentially a human one and not an explication of an aspect of physics.

Back in the first chapter, a distinction was made between circular motion with rotation and circular motion without rotation—the latter, no oxymoron, consisting of motion of an object (or person) in a circle where the object (or person) keeps the same orientation, for instance, facing northward. While that distinction will become important later in this chapter, I think it might be better to begin with yet another way to distinguish circular motions. Here both kinds involve rotation, but they represent rotational modes that differ from each other in a surprisingly basic physical sense.

First, we need a physical quantity, angular momentum, which appeared briefly when talking about flywheels and in a few other places; it now deserves an explicit reminder. Angular velocity is the rate at which some rotating body turns, perhaps given as revolutions per minute or degrees per second. Recall (or bear in mind) that multiplying an object's mass and its speed gives the tendency of something, once on its way, to keep going straight—its linear momentum. Analogously, the tendency of something to keep rotating, its angular momentum, is given by the object's angular velocity and something somehow related to its mass. One obtains this last factor by multiplying each little element of the object's mass by the square of the distance of that little element from the axis about which the body is rotating. One then adds up the results of all those little multiplications, the sum of which is called the "moment of inertia" of the object. Again, that square of the distance in the calculation explains why flywheels do best with their masses concentrated as far from their axes as possible.

When you set something into rotation by giving it a shove, you give it angular momentum, and that momentum keeps it rotating until slowed by whatever frictional forces lurk—or until something provides a counteracting shove in the other direction of turning. No big deal; that's what we've been talking about for the past nine chapters. One can even recognize a physical law, the conservation of angular momentum, which helps analyze a variety of problems and aids in designing a variety of useful things, such as flywheel-assisted devices. Right off the bat, that law tells you why you needed that initial shove. You had to impose a force on something external to the system you were setting into rotation, in effect setting something else into rotation in the opposite direction. Most often you push on something so massive—essentially the earth—that its change in angular velocity passes unnoticed, even undetectable, even if your push remains inescapably evident. So—an initial push with a countervailing push on something external to start your object going around, but sustained motion with no further intervention once it's going.

Here's the oddball aspect. Turning with some angular momentum, usually imparted from elsewhere as described above, may be easy to envision. By contrast, turning with zero angular momentum sounds even more self-contradictory than irrotational circular motion. One occasionally runs into claims that it can't be done, even by an occasional physicist, who might be expected to know better. But zero angular momentum turns ("ZAM-turns," we might say) do happen and even have some utility in nature. All they take is a clever sequence of rearrangements of a body's (perhaps your own body's) distribution of mass—its moment of inertia—again something mentioned earlier in connection with flywheels and just explicitly defined.[1] No push from outside the system need be provided. To remind ourselves, one obtains a body's moment of inertia by multiplying each little bit of its mass by the square of the distance of that bit from the axis of rotation and then adding up all of those separate calculations.[2] Figure 10.1, a set of stick figures, together with its figure legend, shows a simple way to execute a zero angular momentum turn.[3]

No, I don't have an application of ZAM-turning drawn from ancient human technology or even from contemporary technology. But spreading the news of its practicality might stimulate further investigation or even invention. In any case, the biomechanic can't dismiss it as mere physical curiosity since animals have been quite convincingly shown to take advantage of the possibility. The paradigmatic case is that of a falling cat that, whatever its initial orientation, persists in landing with its feet properly downward, as in figure 10.2. The cat takes advantage of a highly flexible backbone and its consequent ability to be either extended straight or bent into a back-to-belly (dorso-ventral) arc. The tail turns out not to be massive enough to be of much importance, so tailless cats (whether the condition is inherited or acquired) do about as well.[4] The legs also play very little role, for much the same reason. Geckos (small climbing lizards) do the trick as well; they do press their tails—long, flexible, and massive ones in their case—into service.[5] Springboard divers may jump upward with a considerable amount of angular momentum, but they clearly can add some ZAM-turning to that. One can imagine arrangements of motors located off center on turntable platforms that by themselves turning steadily make their platforms turn

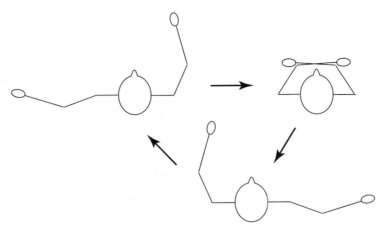

Figure 10.1. How you can do a zero-angular-momentum turn. The figure, a top view, assumes that you're standing on a turntable and shows the motions in your frame of reference (i.e., the resulting whole body rotation is masked). Both of your arms describe counterclockwise circles in a horizontal plane. Holding weights of a few pounds in each hand (increasing their moment of inertia) improves things. Conversely, performance deteriorates if you try this sequence of movements in a swivel chair because your legs are no longer directly under you, which means that you can't mess around with as large a fraction of your angular momentum.

more episodically—and thus avoid either rigid gearing or clutching mechanisms. The appendix provides instructions for a do-it-yourself version of such a device.

A personal experience left me with an indelible sense of another odd aspect of circular, rotational motion. This one is peculiar to one form of so-called non-Newtonian fluid—that is, fluids that aren't just highly viscous but have traces of actual solidity and other complexities. To fix the leaky, nearly flat roof of a shed, I bought several large cans of liquid asphalt roofing compound, the basic version that comes with fibers suspended in the asphalt. Naturally, the components had separated, which demanded vigorous stirring of the all-too-viscous stuff. Yes, stirring with a small paddle did the job, but the

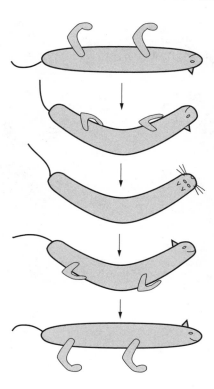

Figure 10.2. My very crude drawing of a cat righting itself as it falls by torsion and flexion of its torso.

muscular burden stimulated efforts to apply external power to the task. So I chucked the shaft of a paint-stirring propeller into my electric drill and off-loaded the task onto the power company. Result— total failure. Spinning the propeller had the perverse effect of, one might say, unstirring the mix. At the same time, black glop crept up the shaft of the propeller. In short, I had my nose rubbed in what's called the "Weissenberg effect" (or sometimes "strangulated flow"). Whether microscopic fibers or long-chain macromolecules, rotation causes elongate inclusions in a fluid to, in effect, wrap themselves around whatever is rotating (fig. 10.3). Stirring without rotation, as when one hand-stirs cake batter or kneads dough, does its job; stirring with rotation of batter or bread dough quite often has this paradoxical outcome.

Figure 10.3. Strangulated flow in a solution of methyl cellulose just after a Dremel tool turning at its lower speed has been hand-lowered until just beneath its surface. For reference, the shaft in the chuck of the tool is ⅛ inch in diameter.

You can experience the phenomena easily enough—the appendix provides several practical suggestions. For that matter, you may have already experienced it, noticing how beating egg whites or some cake batters causes them to creep up the shafts of motorized mixers. Incidentally (lest you waste your time), neither molasses nor pasteurized egg whites will strangulate significantly.

Back again to that distinction between circular motion with and without rotation. Although often quite willing to move in circles, animal appendages remain—as brought up repeatedly here—incapable of continuous rotation. But circular motion without rotation isn't just a subterfuge of the anatomically challenged—it happens more often than most of us imagine in non-living nature. (I almost said "inanimate," but since this account is all about motion, the word subtly misleads.) If you spin a solid body, it unquestionably rotates. But what if you spin something that, while properly endowed with mass, entirely lacks solidity, in short, you spin a liquid or a gas. Now the bits and pieces can move not only as a whole, but they can move with respect to one another. That introduces a world of additional options.

One of these options occurs quite commonly and does so for a reason that's not hard to appreciate. Say a small chunk of liquid or gas—hereafter "fluid"—were to move inward toward the axis of rotation with no change in its angular momentum. It would do, more or less, what a skater's body does when the skater moves arms inward—it would move more rapidly around that axis. As it happens, that increased rate of going around the axis would change its orientation exactly enough to cancel out its rotation. So, as in figure 10.4, it would hold its original orientation as it went around. Vortices in nature behave in just this way, with speeds increasing toward their centers, whether typhoons, hurricanes, tornadoes, the water going down the bathtub drain, or the tip vortices one sometimes sees as vapor trails behind airplane wings. Solids turn with constant angular velocity—speed increases steadily with distance from the axis. Fluids ordinarily turn with constant angular momentum—for these last, speed times the distance from the axis remains constant. At least they do so if free to do so, which means ideally. (The reader of particular percipience or severe skepticism will object that the rule just introduced implies an infinite speed of flow at the axis. No, that doesn't happen—within every irrotational vortex lies a rotational core, the eye of the storm, so to speak. And, for related reasons, the smaller the system, the less of a vortex will be irrotational and the more it will form a rotational core, until one gets down to a domain in which vortices themselves no longer occur.)

An irrotational vortex, meaning the ordinary kind of vortex, may not rotate, but, as just explained, by no means does it lack angular momentum. As a result it shares with rotating solid bodies the normal difficulty of starting up. You can create a close approximation of

(a)   (b)

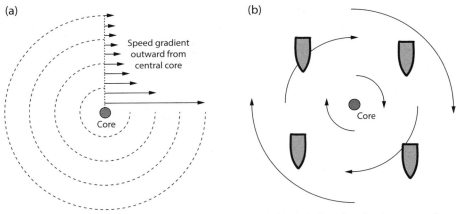

(Arrows indicate local flow speed at their bases)   *(Arrows indicate flow direction, not speed)*

Figure 10.4. Two views of a vortex, fluid structures being hard to reduce to two-dimensionally intuitive form. These are easiest to envision as top views of liquid vortices with vertical axes. In (a) one sees how the speed of flow decreases outward from the central rotational core and how, in a sense, the vortex has no clearly definable outer limit. In (b) one sees the peculiar consequence, how an object commonly makes non-rotational circles as it goes around the core.

an irrotational vortex by stepping out of the system, by, for instance, giving a bowl of water a spin. And, yes, you can create one entirely within the system—except that you simultaneously start another vortex with the same strength of spin (technically the same "vorticity") but going in the opposite direction.[6] Often this other one is so large and slow that it passes unnoticed. Or it can make trouble. To achieve lift, an aircraft wing develops a vortex ring that envelops its length. So, going down the runway a heavy plane (needing lots of lift) leaves behind another of the same strength that can afflict the takeoff of any small plane that follows behind the first—unless local winds blow this starting vortex out of the way or enough time elapses for it to dissipate. You can't see the bound vortex around the wing, but the development of that starting vortex of equal and opposite strength tells us that it's more than some mathematical construct.

In one of its most potent applications, rotary machinery drives propellers or extracts energy from windmills. The functioning of propellers and windmills, both fully rotational as solid appliances, as well as fixed aircraft wings, depends on these bound, irrotational vortices in the adjacent fluid, so the vortices take on at least indirect relevance for the present story. While I have no intention of developing the surprisingly counterintuitive story of how airfoils and hydrofoils operate, I ought to at least comment that the explanation that you've probably heard for how wings generate lift should be disregarded as a polite fiction.[7] This is the explanation in which successive bits of fluid split at the front and rejoin behind an airfoil after taking paths of different lengths and hence going at different speeds above and below. The successful integration of a proper theoretical explanation of the origin of an airfoil's lift with experimental work on airfoil geometry awaited the twentieth century, well after we had a good idea about such things as the basic physics that applied to steam engines.

Waterwheels and paddle wheels depended on maximizing drag, not developing lift in the way of wings, propellers, and turbines, and drag maximization holds no great mystery to anyone familiar with oars or paddles. Unsurprisingly, these were our earliest rotational devices for either extracting power from moving water or applying power to move through water. Until the development of ships' propellers in the mid-nineteenth century, windmills were the only commonly used lift-producing airfoils or hydrofoils. Windmills represented another of the developments of the High Middle Ages, power sources unknown in classical antiquity—yet another reason for the long dominance and persistence of muscle-powered technology. The first good analysis of the operation of a windmill was that of John Smeaton, in 1759, and he treated the fluid mechanics on a purely empirical basis—as well as running afoul of errors from the use of whirling arms, as noted back in chapter 1.[8] The nineteenth-century fabricators of both ships' propellers and windmill blades paid little attention to the critical matter of the contour of their surfaces, as one can see in museums. Even without a decent understanding of the fluid dynamics of lift production, these propellers proved superior to (and largely displaced) side wheels, just as the latter had displaced stern wheels.

A final glance at rotational gear. In this era of digital electronics, dazzlingly complex software, and sealed boxes that operate almost noiselessly, we too easily lose sight of the spectacular complexity of rotational machinery we humans have contrived. Household record changers have disappeared, as have the great Wurlitzer and Seeburg jukeboxes of Main Street diners. Only when glancing over the shoulder of loader or fixer does one glimpse the complex machinery behind a mechanical coin-operated vending machine. Never mind player pianos—over a hundred years ago inventors produced entirely mechanical string quartets. Rube Goldberg now lives on as little more than a poorly appreciated expression.[9]

Prefatory (or preambulatory, one might say) to an imagined technology without any form of continuously rotating machinery, we might have a look at just what we actually do with the existing stuff. Not that the subject hasn't come up earlier. Piston rods and their opposites, pitman arms, combined cranking and reciprocation. Cranks themselves as well as epicyclic gearing did the same for circular translation and true rotation. We've talked at length of wheel-and-axle devices and, to a lesser extent, of pulleys, windlasses, and gear pairs. Among gears, we've seen bevel gears and worm-and-wheel gears, plus their counterparts among the wooden devices of earlier times, crown and lantern gears—all of these turning whatever rotated by 90-degree angles while often changing shaft rotation rates as well.

Now a full exposé of the practical combinations of gears, shafts, and levers won't even be attempted here—a large museum might suffice; a small book shouldn't even try. But we might examine the tip of one iceberg in particular. Consider the wide range of clever little gizmos that make steady rotation into something else and, even more specifically, what happens when the output of an ordinary, constant-speed rotating motor passes through some odd but useful pairs of gear. For a comparative view of what happens, we'll match a drawing of each kind of gearing with a simple graph, with the speed of rotation of the output shaft displayed upward and with time progressing left to right.

Non-round gears—elliptical or square ones. Wheels should be round, not elliptical or square, right? Not always. Two identical, non-round gears can be meshed so long as the distance between their axes of rotation—their shafts—doesn't change as they turn (fig. 10.5). Drive one at a steady rotation rate—what does the other do? It turns in the opposite direction, as any proper gear should. But it speeds up and slows down. If it's elliptical, as in the figure, it does so twice with each full revolution, with the height of the peaks and the depths of the troughs dependent on the degree of ellipticity. If it's square, it speeds up and slows down four times within each full revolution.

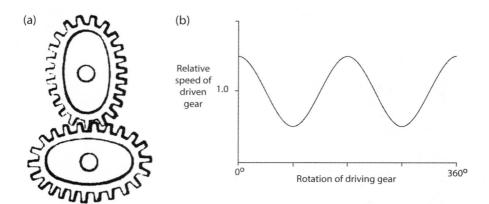

Figure 10.5. (a) A pair of elliptical gears (from Hiscox, 1907), and (b) how the speed of the driven gear varies relative to that of the driving gear as the two turn—at the same (as they must) overall rotation rate.

Just conjure up an image of a bumper car ride at an amusement park where each car has its engine coupled to its wheels through such a pair of gears!

Spiral gear pair. Even that degree of symmetry isn't strictly necessary. Perhaps more practical applications might come to mind for a more eccentric pairing, here one in which the gears are opposite spirals (fig. 10.6). Turning one steadily will cause the other to gradually increase its speed as it makes its revolution, dropping back at the end and beginning again as it turns a second time. Some manufacturing or handling machinery might use it to begin its action with a gentle motion so as to avoid jarring or cracking the objects of its attention. Or it might find use in bringing a mass into motion without imposing a high initial load on a motor, thus allowing a smaller driving system for a weight-limited system.

Mangle wheel and pinion. A step up in complexity gives us a pair of gears that, in response to steady rotation of a small pinion gear, produces an output that rotates slowly in one direction and then returns more rapidly to the original orientation (fig. 10.7). (In another permutation the speeds are equal.) Here provision is needed for one additional element of flexibility—the axes of the two gears have to shift

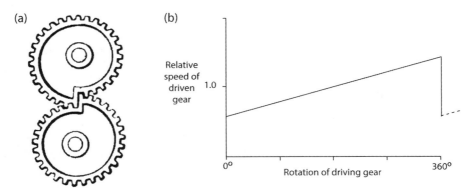

Figure 10.6. (a) A pair of spiral (or scroll) gears, and (b) the ramp up in speed that one can get by driving one of such a pair. Once again, the gears must turn without changing the distance between their shafts.

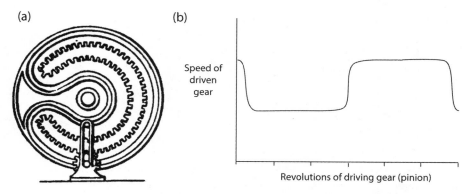

(a)   (b)

Speed of driven gear

Revolutions of driving gear (pinion)

Figure 10.7. A mangle wheel and pinion. The slotted stand allows the pinion to rise and fall, guided by the slot in the main wheel. You should imagine the pinion driven by a belt that runs laterally to a motor, thus permitting that up-and-down movement. The graph ignores the periodic direction change of the wheel.

closer and farther apart as they move. The name "mangle wheel" derives from an old device often used to squeeze water out of washed clothing before drying—a "mangle" in Britain, a "wringer" as on my mother's old washing machine here in the United States. A mangle has two wooden or rubber rollers that rotate in opposite directions, putting the squeeze on the sodden fabric. Driving the mangle with a mangle wheel makes it ingest and then eject the cloth, so, unlike the way my mother's simpler unit worked, one didn't need access to the other side. The mangle wheel and pinion is an old device. I ran across an allusion that assumed familiarity with it in an 1834 patent for an improvement to a cotton-spinning machine.[10] As an aside, the opposite rotations of the two wheels of a mangle could be regarded as canceling and thus producing a mechanism with no net rotation—in a way analogous to the opposite spins of paired vortices, mentioned earlier.

Mutilated gear. This suggestively dysfunctional name refers to the driver of a pair of gears that lacks teeth on all but a small portion of its periphery. Turning steadily, it makes the driven gear turn inter-

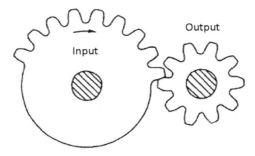

Figure 10.8. A mutilated gear pair for intermittent rotation. In practice, a holding plate on the input and mating sector disk on the output keeps the output gear from freewheeling when not engaged by the teeth of the input.

mittently (fig. 10.8). To appreciate the utility of the pair, imagine that you have an ordinary electric motor, one that rotates continuously, but you want a secondary rotor to move stepwise, in increments of, say, a tenth of a full revolution, or 36 degrees. Why might that be desirable? If the driver turns at a revolution each second, then the driven will turn by a tenth of a revolution every second, perhaps thereby exposing another digit each second. If that second gear has on its shaft a similar mutilated gear, then the next one in the sequence will advance a unit every ten seconds—and on and on. You have a digital readout of the accumulated rotations of a non-digital driver. The water meter on my house depends on just such a mechanism, its motor a little turbine in the water line.

Years ago, banks of mutilated gears nestled within motor-driven digital clocks. Every minute a digit would suddenly move upward, to be replaced by the next; every ten minutes, two digits would move upward; every sixty minutes, three would move, and so forth, repeating every twelve hours. The well-regulated 60-hertz electrical supply established the accuracy of the underlying motors, so these were fine timepieces, gracing the tops of 1950s household television sets.[11] I'm amused to see purely electronic digital clock displays designed to resemble these old gear-train clocks. Their numerals appear to flip upward as one disappears and the next appears—one can't imagine a better example of a technological atavism.

Geneva drives. Unless equipped with some form of holding ring, a lossy frictional kludge when not engaged, the gear driven by a mutilated gear is free to follow its own freewheeling fancy. Furthermore, the sudden engagement of what teeth remain with the now-stationary driven gear heavily stresses those few teeth—too many of us have had too much experience with overstressed teeth, even if mainly dental rather than mechanical. A better, if more costly, way to bring the driven rotor under full control is with a so-called Geneva drive, especially good for higher loads and greater speeds. The name of this ever-so-clever gadget comes from its original use in mechanical watches, for which Geneva, Switzerland, was long renowned. Figure 10.9 shows how one works.[12] Like the mutilated gear, the ratio of driver turns to driven turns is always an integer, one fixed by the geometry of the pair.

Geneva drives were often (and occasionally still are) used in movie projectors, where individual frames had to be quickly brought into the focal plane, held there, and then as quickly replaced by the next—the normal frame rate was (and still is) twenty-four frames per second, and the brightness of the image on the screen depends in part on the fraction of that twenty-fourth of a second that frames are held in the focal plane for illumination.

The day fast approaches when household 3-D printers and downloadable software will make it easy to assemble working models of all these mechanisms. At this point, one has only diagrams and commercial examples for most. So perhaps—lest I lose the reader in a zoo of cams, ratchets, worm gears, pinions, pawls, and rolamites—a few simple rotational devices might be introduced that you might even improvise yourself.

One of real everyday utility—a twist-tightener or twist-vice or rope-twist lever—is too ordinary to gain the attention of most compendia.[13] It puts ever-increasing tension in a loop of rope by twisting

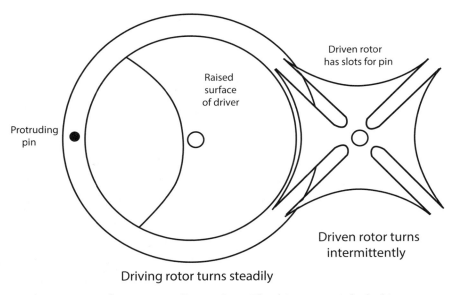

Raised
surface
of driver

Driven rotor
has slots for pin

Protruding
pin

Driven rotor turns
intermittently

Driving rotor turns steadily

Figure 10.9. A four-position Geneva drive. The driven rotor is locked in place as long as it contacts the raised surface of the driver. When rotation of the driving rotor brings the cutout region of the raised surface and the pin around to the driven rotor, the latter is forced to rotate 90 (and only 90) degrees.

it, doing that by rotating a stick inserted in the middle, as in figure 10.10. The rope receives no net twist since half twists one way, the other half the other way, but the stick must be rotated to generate tension, which, not incidentally, can be extremely great with only a modest application of manual force. Best results come if the rope is mechanically stiff—if, more specifically, it strongly resists elongation as the device comes into play. Otherwise more rotation of the stick will be needed and, worse, more elastic energy will be stored up that, upon sudden and untimely release, might damage the operator. This last occurrence can't be blithely ignored—withstanding the surprisingly great force takes a stouter rope than you might initially anticipate using.

Where might one use such a twist-tightener? I think the last

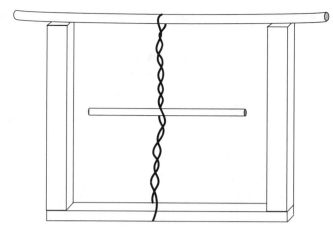

Figure 10.10. A rope-twist vise, here just used as a demonstration of how a sturdy stick (such as a broom handle) can easily be fractured with a bending load.

ones I improvised helped me hold the lower crosspieces of wooden chairs in place while filler and glue slowly set. The loops encircled adjacent legs, and, being just ropes, they produced none of the scars that metal furniture clamps might have left. Long wooden spoons served as tightening sticks, if I recall right. The basic arrangement can do many jobs, large and small; the only odd requirement consists of some way to keep the tightening sticks from rotating back again on their own. A twist-tightener might adjust the tension in the end rope of a hammock. Or with a pair of stout screws for the ends of a diagonally running loop, it might square up a screen door. During construction, a larger version could serve as a temporary squaring system for the frame of a wall. Its main strength imposes its main limitation—a twist-tightener provides a huge force but moves it only a tiny distance.

With my all-too-mechanical prejudice, I suspect allusion in the Old Testament. Samson, in Judges 16:28–30, blinded prisoner of the Philistines, is brought out and restrained between the columns of their temple for the Philistines to torment. By pulling or pushing (literalist biblical scholars argue endlessly about such matters), he crashes the columns and brings the roof down upon the lot, including, of course, himself. Never mind his regrown hair—did Sampson recognize that the Philistines had looped the restraints around adjacent columns and provided a stout pole within reach? So equipped, he might, without vision or superhuman force, have done sufficient twisting either to cause buckling or dislocation of a central element of one or the other columns. While on old allusions, bear in mind that the words "torsion" and "torture" share the same Latin root as well as close similarity in English and modern Romance languages.

The classic application of a twist-tightener tensioned the blade of several permutations of a framed saw, specifically bucksaws and bow saws. Recall yet again that, until fairly recently, economical design meant minimizing the need for metallic components. Iron made good saw blades; steel made even better ones—both were especially expensive in the high-quality forms that could make good saws. Stretching a blade lengthwise during use allowed a thin, narrow blade to cut through a fairly thick piece of wood, and the more forceful the stretch the better (at least short of the tensile limit of the blade). According to reasonable evidence, bucksaws (figure 10.11a) go back to the Romans. They did well as cross-cut saws, but the compression-resisting crosspiece prevented them from doing all but short ripping jobs. Bow saws (figure10.11b) looked almost the same, but their blades could be rotated, pushing the crosspiece over to one side, which permitted long ripping cuts as well as crosscutting.

Usage (linguistic, not craft), though, has blurred the distinction. Bucksaws remained common for cutting firewood in the United States—although with the tensioning done with a turnbuckle instead of with a tensioning cord—until Scandinavian saws with tubular steel bows became common after World War II.[14] These last—marvelously lightweight, inexpensive, and effective for fresh wood up to about four inches in diameter—have become rarer in the past few years. I can only attribute their decline to the besmirching of their reputation by a generation of Oriental rather than Scandinavian

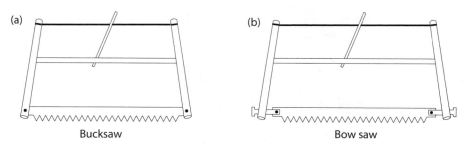

Figure 10.11. Bucksaw versus bow saw; the former is a much older design. The bow saw has a blade that can be rotated around its length, permitting the saw to cut at an angle (up to 90 degrees) to the frame.

blades, superficially similar in appearance, but with ill-set teeth and frustratingly inferior performance.[15] For small jobs and general trimming, gasoline-powered chain saws are more trouble than they're worth next to one of these Scandinavian hand saws.

Twist-tightening appeared in quite a different guise back in chapter 6. The great ballistae of the Greeks and Romans deliberately violated our injunction against using ropes that stored elastic energy, twisting ropes of bovine tendon in order to do just that—to store up the energy for throwing heavy projectiles over long distances. One might perhaps view ballistae as making a virtue of necessity. Since their best available energy-storing material (at least on a per weight basis) had a low extensibility, torsion was one of the few practical ways to load it. And that points up, again, the chief characteristic of this ever-so-simple loading scheme—high force, low distance.

Amplifying the force our muscles can exert matters in a host of situations, so here's another scheme, admittedly one that lends itself less well to quick improvisation. Among rotational devices, windlasses, alluded to in chapter 6, typically hoist things by winding rope onto a shaft or drum. A slight elaboration of a basic windlass circumvents more complex arrangements of pulleys—it's called a "differential windlass" or, less often, a "Chinese windlass" (figure 10.12). Turning its crank unrolls rope from one part of a shaft while simultaneously

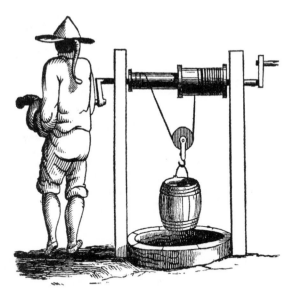

Figure 10.12. A Chinese or differential windlass, a way to get an enormous mechanical advantage at the cost (no free lunch!) of a lot of rope and cranking. This one has a proper pulley rather than merely a greased eye (from Ewbank, 1842).

rolling rope up on another part of a shaft. The key, of course, is that the diameters of the two parts differ; thus if rope is unwound from a narrow diameter and wound onto a wider diameter, the load below has to move upward. If those diameters differ only a little, then there's little upward motion, but what motion happens comes with great force relative to what the crank puts in. Yes, the rope's bottom loop does have to slip through some eye (or pulley) at the bottom, but with a gentle curve and sufficient grease, nothing very sophisticated need be provided.

According to the most accessible source, the term "Chinese windlass" traces to some British soldiers who, near Beijing during the Second Opium War, in 1860, observed a single operator opening a drawbridge with one.[16] In fact, the differential windlass was almost certainly known in Europe at the time,[17] although it was apparently uncommon enough so these observers thought they were viewing an Oriental novelty. But in English usage, the name actually goes back further. Thomas Ewbank, in 1851,[18] mentions it in his remarkable book on the history of water-moving machinery, commenting on how it constitutes yet another of the mechanisms Europeans acquired from the Chinese. On this last point, though, Joseph Needham, a century later, seems less sure. And Needham was the greatest exponent ever of Chinese origins, giving credit not just wherever due but wherever credible.

A huge windlass gave remarkable service as a stump puller—with draft animals pulling in both directions—during the construction of the Erie Canal from the Hudson River to Lake Erie, between 1817 and 1825. But it made no obvious use of the differential trick.[19] And it's a somewhat subtle trick. One first imagines increasing force by pulling a rope off a drum of small diameter while winding a rope onto a coaxial drum of larger diameter—one might guess that the greater the ratio of diameters, the greater the force increase, as with meshing gears. But the differential windlass does best with drums of only slightly different diameters. Think about it. If the diameters were equal, the load would not rise at all. The differential windlass achieves great force amplification while remaining a device that's especially easy to build and use.

What might one gain with a differential windlass? The thing amplifies the applied force by an enormous factor, remarkable for a single, simple device; the price, again, is movement of the load for only a short distance and at a relatively slow speed, at least if the windlass is manually operated. Consider a differential windlass with a larger drum 4 inches in diameter turned by a crank whose circle has a diameter of 32 inches—that's an eightfold amplification. Making the smaller drum is 3.5 inches in diameter, half an inch less than that 4-inch larger one, gives an additional eightfold amplification over that from the crank. So overall, this windlass increases the applied force by a factor of 64! Two people turning cranks at opposite ends could readily generate a full ton of force. Adjustment of diameters and their ratios might coax the output up to two tons, enough to lift one of the stones of the pyramids high enough to place rollers under it—not that we have any reason to believe the device saw such use.

Nowadays hydraulic lifting machinery and hand-operated winches (come-alongs) are cheap and ubiquitous, so one hesitates to claim enough day-to-day utility to offset the bother of making a differential windlass of substantial capability. Still, one can imagine equipping a garage with a differential windlass at one end, a windlass whose cables run over a few pulleys so the business end dangles in the middle—a single person could easily hoist out automobile engine blocks. Or one might build a differential windlass below the crosspiece of a large sawhorse-like stand and use it to pull small stumps or well-seated fence posts directly upward. With some ropes to anchor it to a nearby tree or light pole, the same device might pull a car out of a ditch.

When talking of levers, Archimedes supposedly said, "Give me a place to stand, and I will move the earth." Alternatively, he might have advised use of a rope-twist vise or differential windlass, asking instead for something to which to tie it.

[11]

Rolling Back Rotation

If organisms large enough to for us to see have no truly rotational components, then a mechanical technology without the benefit of rotational devices need not be imagined—it exists, to extend the metaphor, right before (and behind) our eyes. But what might be worth imagining is a version built of components made the way we humans manufacture the components of our technological world. Naturally, to pun slightly, we'd anticipate that this non-rotational, human-built technology would converge with the devices directly generated by the evolutionary process. While biomimetics would still offer no universal magic formula for success, its attractiveness would undoubtedly increase.

Perhaps "imagining" conjures up an unnecessary level of conceptual creativity. After all, however I might disparage the more extreme claims of biomimetics, we do have all of life's devices as models. We might invent legged vehicles, somehow taming their intrinsic complexity, possibly by building leg-tuned road analogs. Beyond such obviously biomimetic solutions, we have a surprisingly rich catalog of things invented and then sidelined in favor of rotational alternatives, devices that might have undergone further technological elaboration were they not judged less promising. Enough, though, of initial generalizations—we have a world to create.

Consider motors. Most of us know only rotational engines, although here and there some unglamorous task receives the output of a non-rotating one—one that moves its business end back and forth in a straight line or in an arc of a circle. Not that both directions need be powered; most often one half-stroke consumes energy directly, producing an output slightly greater than demanded by the load and storing that extra output by stressing a spring or lifting a weight to power the return half-stroke. The straight-line version of an electric motor usually goes by the name "solenoid," providing a nice search term. You start your car with one. Nothing draws as much current from the battery as starting, and no car passes that full current through the ignition switch into which you fit your key. No, turning the key provides a small current for the starter solenoid right next to the starter motor; the starter solenoid, in turn, closes a pair of substantial electrical contacts and activates the starter. (It also makes a gear on the starter motor engage the large flywheel gear of the engine to give the latter its initial spin.)

The starter solenoid is merely the largest and most critical of the linear motors that populate our cars—"relays" we call the rest; they differ only in that energizing the coils of any one of them moves a set of electrical contacts only about a millimeter. The reading head on a computer's hard drive is moved across the spinning disk with a tiny solenoid that turns through an arc. One can be removed from a hard drive that has failed in some other way and made to do other small tasks. Around my lab, a large solenoid (figure 11.1) that once determined whether the drain of our old household dishwasher was open or shut has, over the years, carried out a surprisingly large number of temporary tasks that either had to be done exactly when an electrical signal said to, or else they had to do something in a manually inaccessible place. In essence, a simple electromagnet constitutes a non-rotational electric motor, just one that invests its

Figure 11.1. My ancient solenoid after much use, including (which I do not recommend!) some splashed drops of seawater. Supplying power pulled its core inward (with quite an emphatic bang), which variously triggered cameras, light flashes, and inward yanks on needle-less hypodermic syringes. Ordinary 110 volt AC household electricity powers it.

output in producing a very large force that moves its load a very short distance.

Non-rotational engines, even if we limit ourselves to ones that saw practical service, antedate our present rotational ones. We might discount the ancient steam engine of Hero (or Heron) of Alexandria, which may have piped steam into a sphere, which then spun as the steam came out a pair of downstream-pointing nozzles, as in figure 11.2. Its problem is that it did nothing that mattered, technologically speaking, nor—and this isn't often enough pointed out—was it likely to have had the capability of doing so. The critical matter wasn't the availability of cheap slave labor in the old Roman Empire but the intrinsically miserable inefficiency of the scheme. An aerodynamic or hydrodynamic thruster—whether jet, beating wings, or spinning propeller—operates most efficiently if it throws a lot of fluid rearward at a speed only a little greater than the speed at which the vehicle is moving—or Hero's sphere is spinning. A small,

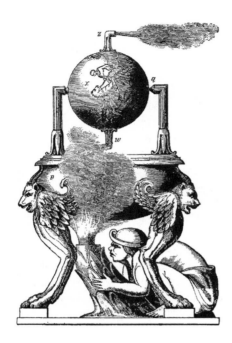

Figure 11.2. A particularly baroque representation of a Hero engine. Note that the artist hasn't been too successful in showing that the jet should be directed perpendicular to the axis of rotation. This illustration forms the frontispiece for the fifth edition of Robert Thurston's history of the steam engine (1895).

high-speed jet in a low-speed application can't achieve any decent energetic efficiency, and that's even before we factor in the losses in coupling it to whatever we want it to power.[1]

Engines of tolerable efficiency at speeds low enough to work with low-precision machinery, the latter built with as little metal as possible, awaited the eighteenth century—and were non-rotational. I won't provide an elaborate history of these early engines, but a few things ought to be mentioned. First, no single inventor should receive full credit. At least three very competent and creative individuals contributed. Around 1700, Denis Papin, a Huguenot Frenchman living for a time in England, made an engine that may have been the first to use a piston. (He also invented the pressure cooker.) About the same time, Thomas Savery, also in England, made a somewhat more satisfactory engine. It used no piston; instead high-pressure steam intermittently forced water out of a closed vessel. But the high-pressure steam proved too much for reliable operation of its

Figure 11.3. A Newcomen engine. Notice that valving was done by hand and that a small pipe was provided to add water above the piston— that helped maintain the seal of piston in cylinder (from Thurston, 1895).

soldered joints and, to be used for pumping, the Savery engine had to be installed down in the pit of a mine. A decade later, Savery joined up with Thomas Newcomen, who produced a fully practical piston-equipped mine pump (figure 11.3) that worked by condensing steam—that is, using atmospheric pressure rather than the steam itself for the push.[2] Newcomen engines gave many decades of good service, especially for coal mines, with their ever-ample fuel supply ensuring that their low efficiency wasn't disabling.

Newcomen engines were huge, slowly operating affairs, with piston diameters of several feet. That size challenged contemporary metalworking, but the low-pressure difference (and allowance for some water to leak in from above the piston) made the crude fit of piston into cylinder tolerable.[3] Modern piston engines trace their ancestry to James Watt's machines. Starting around 1780, oscillating pistons cranked a rotor in these fully rotary steam engines. Watt made a number of other critical innovations—a return to high-pressure steam, adoption of the necessary improved boring technique to make the cylinders, a way to condense spent steam outside the cylinder, and so forth. What needs emphasis, I think, is that, even if Papin, Savery, or Newcomen wanted to build a Watt engine, the technology they had available to them would not have sufficed. For our present exercise in counterfactual history, though, what matters is that non-rotational piston-based steam engines are probably easier to build than rotational ones and might well have been elaborated into a mature technology. And one should bear in mind that Watt's engine was only secondarily rotational, as are nearly all piston engines—pistons normally reciprocate rather than rotate.

A picture of just what might be done with a purely reciprocating steam engine—or how the development of the steam engine might alternatively have progressed—comes from a look at a glorious failure, the steamboat of an American, James Rumsey, of 1787. While its design seems remarkably good (and even more remarkably innovative), it ran afoul of the technological limitations of time and place, and any impetus for its further development was quite likely stymied by the development shortly after it of paddle-wheel steamers driven by ever-better rotational steam engines, starting with the engine Robert Livingston imported from Boulton and Watt, in Britain, for his Hudson River craft.

The engine of Rumsey's boat drew river water (specifically, from the Potomac) into a vertical cylinder beneath a piston through the push of steam introduced above a second piston higher on the same shaft (figure 11.4).[4] Flipping a control valve (automatically) then connected the steam chamber to a condenser; the resultant pressure drop then pumped the water out again. One-way valves (check valves) ensured that water came in from directly below the boat but was expelled through a nozzle at the rear. We'd call this kind of drive a "pulse-jet." Rearward squirting of water, around a thousand times denser than steam, gives a much more efficient push than Hero's engine could achieve. And Rumsey had yet another trick. Along the bottom of the boat, he mounted several additional flapper valves so that the high-speed squirt from the cylinder would entrain additional river water. This increased efficiency by ensuring that the final squirt delivered more volume at a lower velocity; it's the same principle that underlies the wide peripheral chambers around the actual combustion engines in the fan-jets of our present commercial airliners.

Still, I by no means want to argue that, but for what economists sometimes call "lock in," the ships and trains of the nineteenth century would have been driven by non-rotational Rumsey steam engines rather than by rotational Watt engines.[5] The concept of lock-in, which I don't want to disparage as a general idea, might prompt the interpretation that those rotational engines were an accident of the better metallurgical technology available in Britain than in the United States in the late eighteenth century. Once enjoying a certain level of investment and adoption, "lock in" suggests that a technology has the strong advantage of simple incumbency—something starting from scratch has to surmount formidable initial hurdles. The larger, more complex, and more expensive the technology, the greater will be that advantage. No, I think a better case can be made

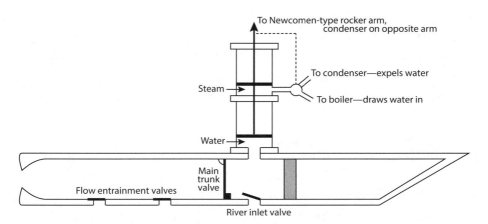

Figure 11.4. The plan of the Rumsey boat's drive. His high-pressure boiler was particularly innovative and probably critical to the adaptation of a bulky Newcomen-like design to shipboard operation. Most of the information underlying the drawing comes from the website of the Rumseian Society, jamesrumsey.org.

that Rumsey's engine (and similar designs considered by his contemporaries) were sufficiently inferior to have been displaced early on.[6] At the very least, the jet-pump design adapts poorly to non-vehicular use, while stationary steam engines enjoyed a huge industrial role early on—far beyond pumping water out of mines.

Non-rotational electric motors abound, but non-rotational combustion engines remain far from the mainstream of current technology—even though, as noted, piston-using combustion engines all had reciprocating first stages. But the fact that the first (or several of the first) electric motors didn't rotate draws far less attention than the space given to those fine Newcomen non-rotational steam pumps. The first electric motors served mainly to show proof of concept and now mainly represent historic curiosities and models for building by students and science-fair folks. Attribution of specific inventions, however sanctified by history, typically contains a large element of arbitrariness. For electric motors, Joseph Henry has come to represent the specific figure most responsible, inasmuch as he was the first person to produce a recognizable piece of hardware. Even he claimed no practical utility for his motor. Henry devised his motor in 1831, when he was a young teacher at Albany Academy, in Albany, New York. He had previously built what was then the world's most powerful electromagnet, one that could support over a ton; he referred to his motor as a "philosophical toy," with no sense of how illustrious would be its progeny.[7] As in figure 11.5, it amounted to no more than a pair of electromagnets, a rocker arm, and some intermittent contacts. Mainly, it worked, it did so in an obvious way, and almost anyone could make one. All this happened well before Henry rose to prominence as the first head of the Smithsonian Institution.

In all fairness, focusing on Joseph Henry does carry a slightly parochial, even chauvinistic, odor. An electromagnetically driven device appears to have been built in 1822 by Peter Barlow, in England; whether it was a proper motor comes down to a definitional issue. And rotational motors quickly eclipsed rocker arms—William Sturgeon, in England in 1832, and Moritz Jacobi, in Prussia in 1834,

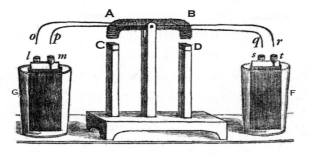

Figure 11.5. Joseph Henry's rocker-arm motor. C and D are magnets, while F and G are batteries. Thus when, for instance, the B-end of coil A-B tips down toward D, q and r contact s and t; that energizes the coil, makes D repel B, and thus makes the arm tip the other way (from Henry, 1831).

produced arguably practical rotational electric motors.[8] One also encounters the not-unreasonable claim that Michael Faraday's invention of the electrical generator, in 1831, makes him at least a co-discoverer. After all, a generator is just a motor in reverse, with mechanical work going in and electricity coming out; on several occasions I've improvised perfectly satisfactory tachometers by connecting tiny unpowered electric motors to voltmeters.

Even now not all of our large-scale motors depend on rotation. Several cities have developed public transport systems that incorporate so-called maglev trains, each car of which is both supported and driven electromagnetically. While expensive to construct, the systems offer relatively low maintenance costs and fine reliability. At present, they operate in Shanghai, China; Nagoya, Japan; and Incheon, South Korea. In essence, a maglev motor is a solenoid, thus a linear motor, but one with a travel measured in tens of miles, not the millimeters or inches of the classical forms.

Other than maglev trains, I think all our motorized ground transport systems turn wheels with their engines. So they can't escape rotation. But one early steam-driven train system had no driven wheels. Its wheels rode on tracks, but the push came via pistons traveling through continuous tubing under the impetus of a steam-generated

pressure difference produced in non-moving power stations. The
pistons connected with the carriages through plates that ran upward through a lengthwise, leather-lined slot in the tubing. Keeping
the internal pressure below (as in a Newcomen engine) rather than
above that of the atmosphere kept the slot normally closed. These
"atmospheric railways," of which I. K. Brunel's South Devon Railway
of 1847 was the best known, turned out to be uncompetitive, in large
part because the leather slot-liner deteriorated rapidly. Too bad—
they provided a far pleasanter experience for passengers than did
contemporary steam trains.[9] Still, they amount to proof of concept—
replace wheels with some alternative bearing system, generate steam
with non-rotational machinery, and one has a non-rotational ground
transport system.

Our familiar jet engines compress incoming air with rotational
turbines, these in turn driven by turbines in the combustion chambers. But two kinds of jet engines dispense with turbines altogether;
admittedly neither is in routine technological use. A wide diversity of
animals such as squid, jellyfish, scallops, dragonfly larvae, and a few
bony fish move around with the first. In nature's version of a pulse-
jet, water enters through one-way check valves and then is forcefully
squirted rearward with the valves closed. In the ones representing
human technology, air enters the upstream end through an array
of one-way check valves (figure 11.6a). It passes into a combustion
chamber, where it mixes with injected fuel and ignites, either from
a sparkplug or from the residual heat of the previous pulse. Most of
the resulting combustion products go out the rear nozzle, with those
check valves preventing regurgitation. The most famous pulse-jets
were the V-1 "buzz bomb" cruise missiles flown by the Germans from
continental Europe to Britain in 1944—noisy, inefficient, inexpensive, unguided explosive carriers. While some forms of pulse-jets
do without physical valves, the high noise levels remain. Ramjets
(figure 11.6b) are the other type of non-rotational jet. Unlike pulse-
jets, which can operate at low speed, ramjets work only at speeds
higher than those common in commercial aviation—optimally at

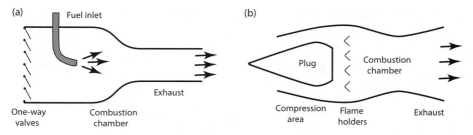

Figure 11.6. Pulse-jet (a) versus ram jet (b). Both may need no continuously rotating components, but their operative bases differ considerably. They do share one other feature—inefficiency.

about three times the speed of sound (thus at about 2,000 miles per hour). Forward motion of the craft compresses the air entering the engine, and careful shaping of intake, combustion, and exhaust chambers ensures that aerodynamic shock waves will direct the output rearward. While technically successful, they have seen very limited application.[10]

Of course, muscle—even if contractile rather than expansionary—sits just as fully in the non-rotational column. Considerable effort has gone into searching for a manufacturable equivalent of muscle, not so much from a misguided assumption of its natural superiority but for all the prostheses that might then be so much closer to the natural function they replace. At this point, no electroactive polymer or other device comes within range of muscle's efficiency, expressed either energetically (output divided by input) or as power-to-weight ratio. And, I hasten to add, that's not to declare that muscle's motor performance is spectacular in either regard, but merely that its specific performance as a contractile motor isn't easy to match.

As a note of comic relief from all this high-speed, high-tech, large-scale technology, consider a non-rotational substitute for an overshot waterwheel that you might build yourself for the amusement of family and neighbors—as I did one Sunday afternoon many years ago. To remind ourselves of its functional essence, an overshot waterwheel,

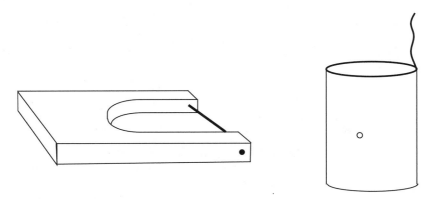

Figure 11.7. The basic unit for the intermittent puller (with string on can) or falling-water mobile (without string). Coffee and soup cans plus supports of wood with either threaded rod, very long nails, or long, thin machine screws do the job. The creativity comes in designing the superstructure and perhaps some water-directing spouts for the cans.

the most familiar kind of waterwheel, extracts mechanical power from a stream of water dropping under the urging of gravity and in the process turning the wheel. Say you have such a source of running water, from hose or rain gutter or small waterfall. All you need as a reciprocating power extractor is a cylindrical bucket that can fill when upright and then invert itself to empty; inversion pulls a cord that goes upward or a rod that goes downward to do the engine's work, as in figure 11.7. Running a rod horizontally through the bucket, just below its center and slightly to one side of the maximal diameter, assures proper tipping. If art takes precedence over work, make a group of such buckets of different sizes arranged so various ones empty into others and where several cycles of one are needed to trigger emptying of another. If you are pathologically compulsive, you might make an array with a sufficient number of elements to form a digital counter or simple calculator.

With only rare exceptions such as atmospheric railways (Google it), maglevs, and ramjets, linear motors are, by definition, incapable of full rotation and instead reciprocate. Almost inevitably, reciprocation requires some secondary device to reduce or, if possible, to eliminate the pulsatile character of the motor's output. That function of whatever is immediately downstream in the power train can be easily overlooked. Consider piston engines. They may depend on external or internal combustion; if the latter, with either spark or compression (diesel) ignition, and either a two-cycle or four-cycle operation. Whatever such details, some form of crank arm causes a crankshaft to turn. And the angular momentum of that crankshaft plus any flywheel and other spinning stuff buffers the pulsations of the pistons' periodic pushes. We help matters by making the power strokes of multi-cylinder engines happen out of phase—in no automobile do the cylinders fire simultaneously. Old steam engines (except on trains and cars) had large and conspicuous external flywheels. Pulse-jets, especially crude motors, depend on the momentum of engine, craft, and payload to keep going between explosions.

A general point lurks here. Reciprocating motors usually do best with some form of short-term, efficiently reversible, energy storage. So a technology powered by such motors will give more attention to practical ways to manage storage than does our largely rotational technology. But even our present technology has lots of short-term storage schemes, some widely used, some awaiting economic preference, some operating over fractions of a second, some over perhaps a year. So what would change in a non-rotating technology comes down to the degree of dependence on such energy storage; we needn't seek some complete novelty (not that one wouldn't be nice to have). At this point, glance at the nearest bit of electronics that's powered by a rechargeable battery, assuming you're not already reading this on one of them. And then to draw attention to the possibilities, consider a few others—besides the flywheels and rechargeable batteries already mentioned. With no implied significance, I'll put them in a form of temporal order, from the longest to the shortest time period over which they store energy in their most common applications.

Imagine you control the temperature of your home with a heat pump, an oversized refrigerator that moves heat from inside to outside (like a refrigerator) during the summer and from outside to inside during the winter. This particular heat pump has its outside exchanger buried in your yard. So you store up heat in the ground during the summer that will then be recycled to reduce the cost of heating during the winter; you cool the ground during the winter, reducing the cost of cooling during the summer. Not a free lunch, just temporary heat storage, done by taking advantage of the heat capacity of soil and allowing a temperature rise in the summer, a rise that's drawn down in the winter.

Consider a power company that either has a nuclear plant that it prefers to run near capacity all the time (the most economical way to run such plants) or a hydrocarbon-powered plant that's not quite sufficient for the demand placed on it at peak hours. But the company does have access to a local mountain. So during the low-demand hours, it turns on some big pumps and fills a reservoir on the mountain; during high-demand periods, it empties the reservoir through power-generating turbines. The process, operating here and there even now, is called "pumped storage." Here storage is gravitational, taking advantage of water's density and lifting it periodically—typically daily.

Since your nicely rotational bicycle is broken, you're forced to run, accelerating and decelerating one leg after another as well as pushing the ground rearward. (If you run eastward, you ever so slightly slow the earth's west-to-east rotation—momentum insists on being conserved.) Legs have mass, much of it located well outboard from the rotational axes, mainly your hips—although the best runners, such as pronghorn antelopes, have remarkably slender lower legs. But at least you, or most other mammals, minimize that useless accelerational work by storing some of it during deceleration for use in the subsequent reacceleration. When you run, the storage mode is elastic—tendons absorb a lot of force as they stretch slightly in length; bone and muscle make only negligible contributions.

A stream of potable water descends alongside your vacation cottage. You might put a small tank in the attic and keep it full with an intermittently operating pump—gravitational energy storage, pumped storage writ small. But instead you build or buy a wonderful piece of 200-year-old technology, a hydraulic ram, as shown diagrammatically in figure 11.8.[11] You're now independent of electrical supply or generator. Water enters an inlet that faces upstream, flows past an open check valve, and out into the stream a little ways downstream. But the flow itself makes the check valve almost immediately slam shut. That stops the flow—almost. Water is massy and nearly incompressible stuff, so stopping flow produces the banging (water hammer) that we sometimes hear in our pipes when turning off a faucet or washing machine. A hydraulic ram uses that sudden impulse—an upward-directed pipe lets a small amount of the water relieve the sudden pressure increase by flowing upward, here into your attic tank. Of course, with little flow in the main pipe, the check

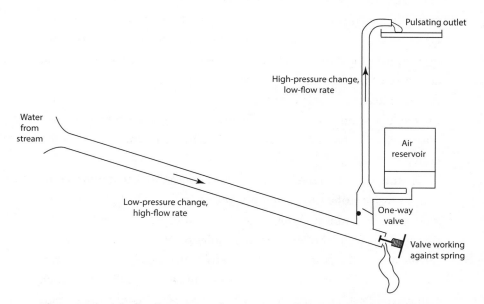

Figure 11.8. A hydraulic ram, put diagrammatically—with (quite abnormally) both valves open at the same time. (If you don't see why this isn't proper, then think again about how the contraption works.)

valve (loaded with a weight or spring) opens again, and the cycle repeats. For efficient and reliable operation, the device needs some cycle-to-cycle buffering, which is provided by the simplest of means, an air-filled chamber, using the compression of the air for elastic energy storage.

Hydraulic rams gave good service on farms located in sparsely populated, hilly areas. A classic (and still very funny) book about life on a chicken farm on the Olympic Peninsula of Washington, *The Egg and I*, makes reference (reverently) at several points to "rams"— undoubtedly puzzling even most readers of the late '40s, when it appeared, and mysterious to almost everyone who might read the book these days.[12]

One last example, from yet another domain. Our lives are entangled in those little black boxes that stick limpet-like onto our electrical outlets and send their outputs to many of our small electrical gizmos and rechargers. Each has a label announcing its input—perhaps 100–220 VAC (volts alternating current) at 50–60 Hz (Hertz, or cycles per second) input, and 5 VDC (volts direct current) output. Ignoring the shift in voltage as someone else's department, consider how alternating current, what the electric company supplies, can be converted to direct current, what most electronic equipment prefers. Our alternating current shifts 120 times each second from sending current in one direction to the other; the average current comes out to exactly zero. So no simple averaging component will do the conversion trick.

A simple check valve does the job, a component that allows current to flow in one direction only. So, as in figure 11.9, the original sine wave becomes a chopped series of unidirectional pulses, sixty per second. Still, the job remains incomplete, with gaps between pulses and voltage varying within each pulse. One can do better, again as in the figure, with a set of four check valves, directing each half wave so each has the right polarity. That eliminates the gaps and minimizes the pulsations. If powering audio equipment, the hum

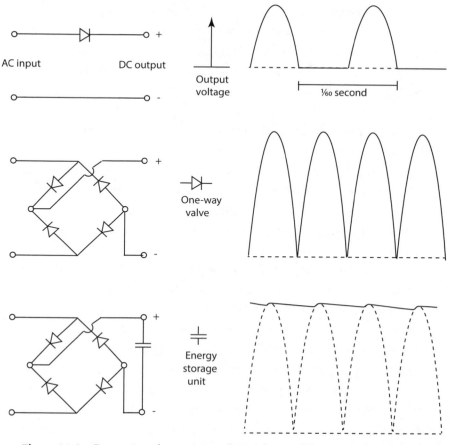

Figure 11.9. Converting alternating to direct electrical current—three stages of complexity. A simple check valve (rectifier) will do, but a set of four (a bridge) does better; brief pulse-to-pulse storage in a very short-term battery (a capacitor) helps still more.

will be less, but it will now be at 120 Hz, which (unfortunately) we hear better than the earlier 60 Hz. What's needed, once again, is brief energy storage. For electronic equipment, the component that does the job is ordinarily a so-called capacitor, something almost indistinguishable from a rechargeable battery. It charges up when the voltage is high, discharging when the load draws on the system. If the load isn't too severe the system of check valves (rectifiers) and storage units (capacitors) can supply a virtually unchanging DC voltage.

Notice how many different modes of short-term energy storage we have available for smoothing pulsatile motion. We only touched on thermal storage, something that can exploit shifts between liquid and gas, solid and liquid, and other effects, besides mere changes in temperature. Chemical or electrical storage, really versions of the same thing, give good service over a wide temporal range. And as additional options, we have inertial, gravitational, and elastic energy storage. But each of these last three—parts of the present story in a way that heat, chemistry, and electricity are not—carries serious drawbacks.

Inertial and gravitational storage require mass. They live and die by Newton's second law of motion and Newton's law of universal gravitation, respectively, the two laws that, for practical purposes, define what we mean by mass. So the whole world of reciprocating machinery tickles a general problem that's largely circumvented by rotational equipment. Mass can be a massive nuisance. It has to be brought into motion (against inertia), hauled around (against friction), and lifted from the earth (against gravity), all costly processes. Inertial storage becomes particularly problematic in a world without flywheels, although it isn't entirely ruled out.

Nor is gravitational storage all that versatile. When we walk, gravitational storage plays a role equivalent to that of the elastic storage of running, but gravitational storage bumps into a speed limit set by the length of the legs of the walker. That's why we have to switch gaits and turn to a different storage mode.[13] The point is general— gravitational storage works only for relatively low rates of reciprocation. And that limits its use mainly to small-scale systems, which tend to be run at higher rates—the great medieval counterweight trebuchets were exceptional in this odd aspect. It also asks for additional linkage elements if the motion is not vertical—when walking, your center of gravity moves up and down, as it must for gravitational energy storage, even if you've no use for anything but forward motion.

Elastic energy storage has its own problems. One needs an elastic, meaning a spring. Which, of course, we have in all shapes, sizes, and relative stiffnesses. We isolate the wheels and brakes of our cars from the rest of the vehicles with springs, and we do lots of other things with them. Organisms do all sorts of things with springs, from saving energy when mammals run or hop to storing the energy with which fleas, leafhoppers, and locusts jump and with which some plants shoot seeds and some fungi shoot off clusters of spores. Most of these applications, though, are, one might say, single-shot games. A spring that returns 90 percent of the energy put in does just fine. But think for a moment about where that other 10 percent goes. In obedience to basic thermodynamics, it appears as heat. Repeatedly stretch and release a thick rubber band, and notice that it warms up, as do the tires on your car after driving for a while at any substantial speed. The most resilient elastomer in nature, resilin, occurs in the wing hinges of insects, some of which beat their wings several hundred times each second. It has been suggested that while 90 percent would be good enough from an energetic point of view, local heating would be intolerable were the resilience not in the 95 percent–plus range. And the bigger the chunk of resilient material, the trickier becomes the problem of dissipating the heat.[14] Compressing and expanding air or any other gas as an elastic may be mechanically awkward, but it does have a real advantage in its minimal time-averaged heat production.

Bottom line, then, for motors and their immediate applications in a non-rotational technology—solenoid-type electric motors would proliferate, along with direct-drive piston engines, so we would still have both electric and combustion engines even if an unfamiliar array of them. We might develop ground transportation based on low-speed pulse-jets, which, for reasonable efficiency, means ones that have large volumes of air going in and combustion products plus air coming out. Perhaps these might operate at frequencies above or below the range of human hearing. Without depending on wheels for traction, vehicles might instead move on well-lubricated surfaces or over other surfaces on air cushions, perhaps with the surfaces contoured for directional control. Maglev systems would more likely do the job for higher speeds and longer distances. Legged vehicles? Even this biologist remains skeptical, impressed by their complexity and comparative inefficiency on any reasonably hard, smooth, and level surface—roads, tracks, troughs.

Much more attention would focus on elastic energy storage. Gravitational storage lacks its versatility. We're stuck with Mother Earth and the terrestrial acceleration of gravity of 9.8 meters per second squared; by contrast, with elastic storage the stiffness of the spring can be altered over a huge range. Still, even with good materials and design, the generation of heat arising from the necessarily imperfect resilience of real elastic materials may be minimized, but it can't be eliminated. So we'd be stuck with another necessary evil and need something analogous to the radiators of our present cars. (Recall that these latter serve only to offload heat that we would also have preferred not to have generated in the first place.) Mechanical devices would as often as possible be contrived so they could operate at the natural oscillatory frequencies of their moving parts—the way a weight on the end of a rubber band has a preferred bouncing speed. Concomitantly, designers would put great effort into minimizing moving mass, which would now have to be accelerated and decelerated, in order to raise that intrinsic frequency.

Also, as implied already, great effort would go into minimizing the pulsatile feel of vehicles—whether ships, planes, trains, buses, or cars—that are powered by reciprocating drives. The drives of ships might converge with those of the fastest of fish, the stiff-bodied, wide-tailed tunas and fast sharks. Still, I doubt if aircraft would find beating wings of the insect or even bird type directly applicable. The preconditions of fast, efficient flight on our human size-scale drove the design of aircraft that separated large fixed wings from small rotating propellers; a beating rather than rotating propeller would not alter the logic. I once saw a video of a tiny demonstration aircraft with a relatively large "heaving" (moving up and down) airfoil as the driver, doing its thing just behind a fixed wing; the design performed impressively well at low speeds. But for optimum performance, larger size corresponds to higher speed, and that still calls for smaller, faster-moving driver units and thus that wing-propeller distinction. Helicopters and Osprey aircraft remain notably inefficient.

One can continue with this exercise in either counterfactual history or science fiction—call it whichever you prefer. Perhaps I should not tickle the world of manufacturing techniques, joinery, and more complex mechanical linkages but just end with a technology that the reader might then extrapolate at will. Before digital timepieces became the norm, rotational machinery ran our clocks. Typically the clock had an escapement mechanism that steadily metered out energy stored in raised weights or coiled springs, or the clock had at its core a motor whose speed accurately followed the 50 or 60 cycles per second alternation of household electricity. A few fancy versions substituted vibrating crystals or some other constant-frequency sources, and some (as in those Telechron clocks of the last chapter) displayed the time in digital form. But all were chockablock full of rotating parts.

Nonetheless, the ancients produced functional non-rotational clocks—mechanical clocks with rotating gears and hands did not grace the bedsides or sideboards or town squares of antiquity. Escapements capable of controlling clocks evolved only around 1300, during that late medieval flush of mechanical ingenuity that came in for so much attention in earlier chapters.[15] So what preceded mechanical clocks? Sundials and sky-watching long antedate any proper historical record, but both demand a clear sky, both have complex seasonal variations, and neither can claim great precision (that is, repeatability) or freedom from systematic (that is, calibration) error.

In at least Eurasian antiquity, a fundamentally different and

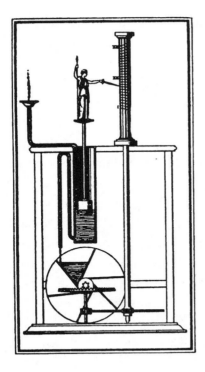

Figure 11.10. A clepsydra, possibly a design of Ctesibus of Alexandria of about 140 BCE. This one is worthy of Rube Goldberg. Siphoned output water hits a horizontal waterwheel that rotates an indicator drum— besides the normal float-based height indicator (from Brearly's 1919 book on the history of time-telling).

widely employed clock that at least eliminated the clear sky requirement monitored the passing of time. Water will drip steadily from a small orifice at the bottom of a container of water; monitoring either the accumulated drip volume or the loss of volume from the container thus indicates elapsed time. The most common generic name for such a water clock comes from the Greek, "to steal," as in kleptomaniac, and "water," as in hydrology—"clepsydra"; figure 11.10 gives one design. Unfortunately, several all too intrinsic problems bedevil clepsydrae. First, as the water level in the container drops, the pressure persuading water to drip also drops and so, therefore, does the volume per unit time coming out. This last variable decreases in a mathematically obscure way not specifically known to the ancients. At the very least the gradual decrease eliminates any easy way to make a container that could have on its side an accurate linear scale.

Second, the downward push of gravity may be dependable, but the counterbalancing resistance comes mainly from the viscosity of water, and water's viscosity happens to be an especially temperature-dependent physical property. For instance, dropping the temperature from 86° to 66°F increases water's viscosity by nearly 30 percent.[16]

Clepsydrae are fundamentally non-rotating, even if with a string and pulley a descending float can be made to turn a dial—as did one Isaac Newton built as a boy. Mechanical clocks that start with an escapement appear to have been rotational ever since their origin—even though escapements need not be. I don't see any fundamental reason why a mechanical, escapement-controlled but non-rotating clock can't be arranged to kick a moving element one unit of travel for each tick, with a second element moving one unit every time the first has gone, say, twelve units. The second unit would at the same time allow the first to drop back to the null position. And so forth. I'll leave the matter here for the reader who's both more compulsive and mechanically creative than the author.

The author's discipline—even if the word "discipline" conveys a sense of restraint that in his case may be inappropriate—is biomechanics, and many of us who chase biomechanics seem to catch biomimetics as well. So I want to end with a reiterative statement on the extent to which a non-rotational human-built technology will, simply by following its own internal logic, converge upon (or borrow from) the devices of the living world. Convergence between the two technologies has come up repeatedly, and some general assertion of its inevitability and ubiquity would be the icing on the cake for the biomimetically persuaded. I do think a non-rotational mechanical technology would look somewhat more like what we see in living nature, but I have my doubts that truly large-scale equivalence of mechanical devices would be likely. The two technologies simply do too many other things differently besides the difference in application of rotational motion. Nature builds big things cell by cell, not by fabricating large parts for final assembly. She uses no metallic materials; we find metals of the greatest utility. Nor does Nature use any engines that extract energy from differences in temperature—heat engines. Her devices tend to be flexible; ours are stiffer. Few of her swimmers go about on the surface of the water; almost all of ours do. And on and on.[17]

But . . . if things go back and forth, then flexible materials have especial advantages, particularly in propelling vehicles through air and water. Brief energy storage, as noted, becomes highly advantageous, additional impetus for the use of elastomeric materials. Our pulse-jets may work by successive explosions of hydrocarbons, but all jet-propelled organisms except single-shot squirters depend on pulse-jetting. So some degree of convergence seems inevitable, along with greater opportunities to profit by emulating nature. That's obviously possible with our present rotational technology, but it's less likely to yield competitively superior outcomes on the large-scale, macroscopic level we've been considering.

In the end, though, one ought to take this highly selective romp through the history of muscle-powered technology as an object lesson in the advantages—especially simplicity and efficiency—of truly rotational machinery. Such advantages long ago justified the awkwardness of powering true rotation with an engine incapable of such motion. Waterwheels, windmills, ships' wheels and propellers, aircraft with decent prospect of leaving the ground—all took up rotation at some very early stage, however "unnatural" the activity. It's just too easy to convert non-rotating to rotating circular motion, if you must start with the first.

[Appendix]

Making Models

Enough talk about all those anachronistic bits of muscle-powered human technology. Direct hands-on experience transcends any number of words—"hands on" in the most direct sense. Actually "hands on" aids understanding and appreciation in two ways—both making the devices and putting them to use. The preceding chapters described quite a few buildable models; the present one adds others that would have unduly disrupted the narrative if given in sufficient detail earlier. I claim no great originality for the models that follow. Perhaps paradoxically, given our world of virtual realities, websites such as YouTube have empowered a subculture of mechanical antiquarians—its attractions are that one can both show off one's cleverness and initiate contact with like-minded fanatics.

I should emphasize, even if overstating the obvious, that these are functional as opposed to structural models, models that work like rather than try to look like ancient mechanisms. In part that reflects my basic outlook as physiologist rather than morphologist—my focus on function over structure. But more importantly, a functional model represents a hypothesis, an idea of how something worked. If your model does work, then perhaps you're on the right track toward understanding something for which you have only an image or a few words of description or allusion. Conversely, some of the most instructive models turn out to be ones that fight back, ones that fail in ways that force you to rethink your initial notions. Either way, you gain appreciation of the cleverness of our forebears and of the limitations under which they labored.

The game is particularly engaging if you're a woodworker—for the most part we're going back to an age when wood provided the main material for mechanical devices. The particularly adept wood-worker will be pleased to enter a world that relied more on wood-working skills, one with few metallic shortcuts such as nails, screws, and brackets. But we'll assume modern tools—making and then de-veloping facility with the tools that once made the items described here would take altogether too much time and effort. (And that as-sumes we know how the tools were made and used.) Still, we'll not assume access to that most wonderful but as yet far-from-ubiquitous tool, the 3-D printer.

Materials. Function remains the main issue here, so we'll be op-portunistic in choice of materials as well as tools. Wood may be the default, but nothing precludes taking advantage of the metals and plastics available in contemporary emporia. These sources provide the real criterion of present-era practicality—as far as possible, the models have been designed around what's on the shelves of hardware stores and other retail establishments. Beyond that, oc-casional recourse has been made to online suppliers. Among the latter, Edmund Scientific (now called Scientifics Direct) comes first to mind as perhaps oldest and largest hobbyist-oriented source (www.scientificsonline.com), but many other useful sources exist, such as

- AmazonSupply, formerly Small Parts, Inc.
  (www.amazonsupply.com)
- American Science & Surplus (www.sciplus.com)

- McMaster-Carr (www.mcmaster.com)
- Woodworker's Supply (www.woodworker.com)

A general (but still limited) directory of suppliers can be found at www.nostalgickitscentral.com/info/parts.html. Mainly, though, just search for whatever particular thing you're after.

Many of us have accumulated odd bits and pieces, reflecting the passion of the mechanically minded but impecunious, immersed in the jetsam of an industrial age. Put less pretentiously, we hate to discard items of imaginable reuse. While my particularly idiosyncratic junk pile (to use a less euphemistic name) may have helped me in prototyping, no design fully reliant on it will be included here. Similarly, I'll try to be as well-behaved when it comes to tools, bearing in mind that my immediate access to a metal lathe and milling machine represents a privilege available to all too few of us. Still, I urge you to think outside the commercial box, meaning the items' ordinary labels. For instance, replacement parts for sliding glass and screen doors represent a source of inexpensive pulleys and bearings, and replacement grinding wheels make good flywheels.

Working models of the larger wooden machines of the ancients should be well within the capabilities of any woodworker with sufficient space, perhaps a good-size personal workshop or garage. But I've limited myself here to portable items, ones that can be carried around to classrooms or other places where one might want to demonstrate (= show off) without special assistance in transportation and reassembly. I've resisted the temptation to make a large whim or treadwheel to replace the treadmill on which I do my regular exercise.

So, ordered according to the preceding text, contraptions to consider constructing, diverse demonstrations to develop . . .

Chapter 1: Circling Bodies

Epicyclic gearing. The Ramelli reader (fig.1.8) almost certainly represents an unreasonably ambitious project, even if it might be quite

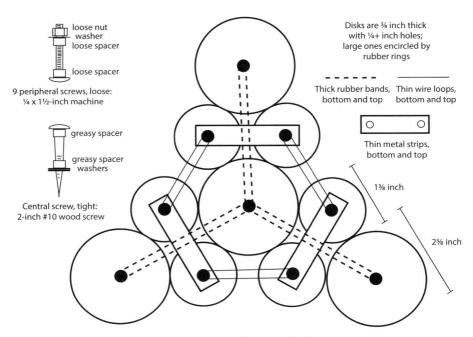

Figure A.1. Epicyclic gearing based on friction wheels rather than gears—impractical, perhaps, but a decent demonstration of the principle.

a fine addition to the library of a retirement community in which many residents are wheelchair-bound. And making gears tooth by tooth is not for the faint of heart. Still, a fair model of such a scheme can be made from some circular pieces of wood planking or plywood (or plastic) around some of which rubber rings have been stretched, as in figure A.1.

For the outer planets to circle non-rotationally about a fixed sun gear, they must be the same size as the sun gear; the size of the inner planets doesn't matter—the present sizes correspond to hole cutters I had at hand. If you happen to make the outer and central disks the same 2⅜ inch, rubber rings can be made as 1-inch circles cut from an inner tube fitting bicycle tires between 1.75 and 2.125 inches. For the metal strips, I used ¾-inch galvanized pipe strap because I had a roll and using it saved drilling or punching holes for the screws. I

screwed the whole thing (via the center screw) down to a square of
1 × 12 shelving board, again because that was at hand. To me the way
the three outer planets walk around without rotating remains counterintuitive. One can demonstrate the principle with only a single
outer planet rather than the three shown here—the three units here
operate independently, with the wires between pairs of inner planets
serving as spacers to offset any slippage.

Chapter 2: Wheels and Wagons

Treenail (or trunnel) joinery (fig. A.2). While slightly peripheral to
the present story, we need reminding of the practicality of pure wood
joinery; in some ways, such as distributing stresses, it can be superior
to casual use of nails and screws. It not only kept buildings standing, but for many centuries held together all the treadwheels, drum
pumps, and the like of the present story. It even allowed wooden
ships to cross stormy seas on a regular basis. You'll be surprised at
how well even a single treenail can hold two planks together.

    As a demonstration, you might use two pieces of 1 × 8 planking

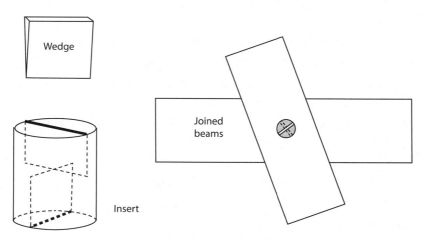

Figure A.2. A treenail joint that's remarkably sturdy and, if sanded smooth,
rather pretty as well.

(really about ¾ inch thick), each about 15 inches long, plus about 2 inches of 1-inch diameter wooden dowel with two 1-inch-wide tapered wooden shims for wedges. Drill 1-inch holes through the centers of each plank, as nearly crosswise as you can. Fit a length of dowel (cut to the right length) into the combined holes. Then remove the dowel and slit each end with a thin saw, cutting a bit over halfway through; the two cuts should be 90 degrees to each other. Insert dowel, hold boards together, and then, one at a time, insert shims and pound them in as far as possible. Cut off the ends of the shims.

Wheeling on soft substrata. Wheeled luggage is now ubiquitous, so you can load up one (the smaller the wheels are, the more persuasive the influence of the substratum) and monitor the force it takes to pull it across floors of varying compliance—in practice carpet of different heights and with different backing thicknesses.

Measuring the force takes a little more creativity, unless you happen to have an old-fashioned hanging spring scale of appropriate capacity. The greater the force needed, the more a piece of luggage will tilt forward; thus you can look at the angle between the long axis of the luggage (marked on it) and a vertical line (a hanging weight). But that assumes some uniform or known distribution of the weight within it. Alternatively, any syringe-like device such as one used to inject basting into roasting poultry can be made into a force meter by inserting a rubber band or coil spring into the space below the piston (fig. A.3). (Large plastic hypodermic syringes, less their needles, work well but may not be easy to obtain.) An internal rubber band produces a tension gauge; a coil spring makes a compression tester. If you want it to record maximum force, grease the inside with petroleum jelly and look at the mark left by the piston after pulling or pushing the suitcase. Pulling or pushing at a steady speed is necessary; how steady depends on the quality of the data you want. But hand-pulling or -pushing will most likely be good enough.

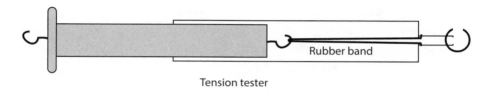

Tension tester

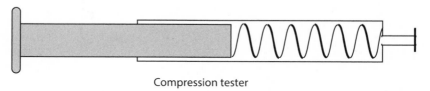

Compression tester

Figure A.3. Two simple force-measuring devices useful for either instantaneous or (with a thin layer of petroleum jelly inside the barrel) maximum indication.

A *Chinese-style wheelbarrow* (fig. A.4). This may be a bit elaborate, but it yields a functional addition to a lawn-and-garden armamentarium. The critical component is a wheel of large diameter, with the front wheel of an old bicycle the most readily available form. The wheelbarrow needs to incorporate provision for removing the wheel without taking everything else apart, and the side compartments should not come too close to the ground, especially on the sides and in front if the vehicle is to negotiate the big bumps that the large wheel accommodates.

Chapter 3: Turning Points—and Pots

Pin bearing (fig. A.5). It's worth trying one of these just to see how easily it's made and how well it works. The main thing for a quick demonstration is to keep the center of gravity below the location of the pin, the latter most likely a wood screw, sharpened machine screw, or nail. If the diametric piece (the axle on which the adjacent

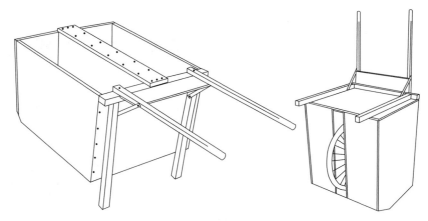

Figure A.4. One design of a Chinese wheelbarrow, with ½-inch plywood as the basic material. A pair of replacement handles for a small conventional wheelbarrow represents a good splurge if you intend to make regular use of the thing. Also, resist any urge to put a crosspiece between the handles as on most yard carts. Such ergonomic horrors prevent the user from bearing the load directly through the long axis of the body, with resulting load diminution and post facto backaches.

part turns) is sufficiently thin, then the device can form part of a drag-measuring gauge or else a deflection anemometer. A slightly V- or U-shaped diametric helps keep parts centered. (A knife-edge is an alternative to a thin, diametric shaft; many commercial balances use just that.) With things balanced on that crosspiece whose drag depends on orientation, one can make neat mobiles—secondary pin-bearing beams can hang on the ends of the main one.

Other bearings. Mechanical bearings can be salvaged or purchased. (I got a fine set of tapered roller bearings from an auto supply store, ones intended for the front wheels of a rear-drive car, shown in fig. 3.3.) But a more interesting pair of bearings can be exposed by dissecting the muscles and tendons off a lamb leg, bought from the nearest meat market. (The muscles then make a fine stew, or one

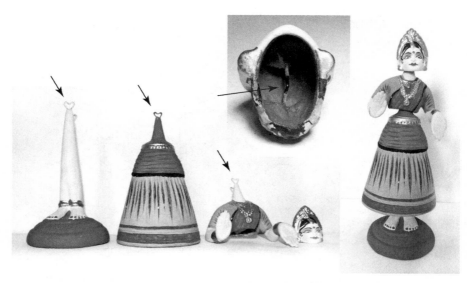

Figure A.5. The Indian dancing doll here has three pin bearings, each biased with a slight bend so she faces generally forward. Each is above the relevant center of gravity, so the doll may move nicely but remains stable. The inset looks at the inside of the head from beneath, with the arrow pointing to the V-shaped bent wire bearing.

can, if adept at the dissection, serve a butterflied leg of lamb.) One obtains a hip and a knee, thus a ball-in-socket joint and a hinge joint. The friction should be impressively low. (Porcine joints will also work, but the leg bones between them are shorter, so the demonstration is less impressive.)

Thrown ceramics. Not everyone has clay in the backyard, and, in any case, no ordinary fireplace or oven will fuse it into a proper ceramic. But you can make a mix that can be "thrown" on a wheel, solidified by baking (try 350°F), and eaten to assay its toughness. A mix of wheat bran and either egg white or egg substitute (Egg Beaters or other brand) should be made up to a consistency just solid enough to retain a shape. One also needs to improvise a wheel. This can be

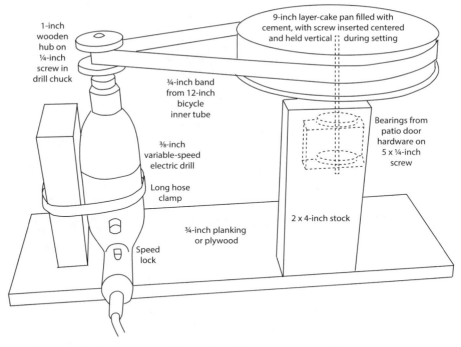

Figure A.6. The improvised slow wheel for turning our edible containers. A few opportunistically positioned brackets have been omitted. Some non-verticality of the side of the cake pan can be offset by tilting the drill. Another source of good bearings: skateboard replacement wheels.

done with a layer cake pan that has nearly vertical side walls filled with mason's cement (to increase the moment of inertia) and belt driven (at low speed!) by an electric drill, as in figure A.6.

Crack propagation. Sharp corners are points at which dangerously high levels of stress can build up, a point that turns out to be re-markably simple to demonstrate—you need only scissors and paper punch, as in figure A.7. A slit in the folded edge of a piece of alumi-num foil extends with only the slightest of pulls; removing more foil with a paper punch and rounding the tip makes it (a bit para-doxically) much more resistant to further extension. The behavior

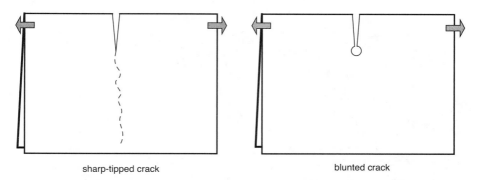

sharp-tipped crack   blunted crack

Figure A.7. Demonstrating crack propagation. Rounding the tip of the crack greatly increases the resistance to elongation of the crack under a tensile stress.

of aluminum foil contrasts sharply with stretchy stuff such as transparent food covering film (Saran or other polyethylene food wraps), where the tips of cracks round themselves under load. Paper is in between—it has the advantage of contained fibers (as a composite material) but with the minimal extensibility of metal foil.

Chapter 4: Going in Circles

Crown-and-lantern gear pair. Making a functioning wooden crown-and-lantern gear pair (fig. A.8) takes a bit of playing around—rewarded by recognizing some subtleties that must have been mundane to woodworkers half a millennium ago. You also get a sense of the way the design neatly accommodates the graininess (anisotropy) of wood. The diagrams give my particular version, one that has a 4:1 step-down gear ratio; the particular dimensions partly reflect available material. Decent meshing depends on tapering the pegs of the crown and matching the spacing of those pegs with the spacing of the lantern's dowels, the last in linear, not angular units. Ease of operation turns out to be humidity-dependent, pointing up a less-often-noted advantage of metallic over wooden gearing.

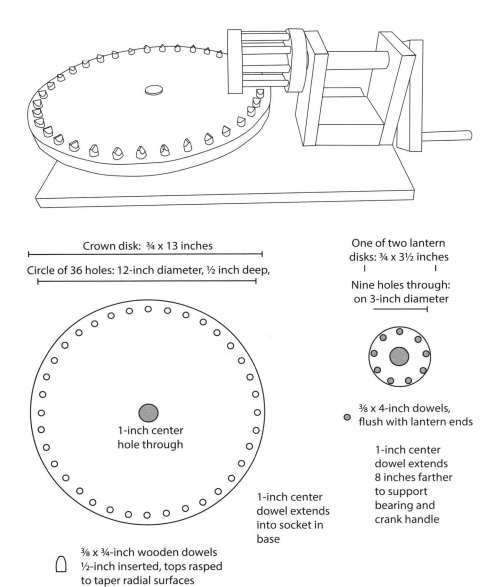

Crown disk: ¾ x 13 inches

Circle of 36 holes: 12-inch diameter, ½ inch deep,

One of two lantern disks: ¾ x 3½ inches

Nine holes through: on 3-inch diameter

1-inch center hole through

⅜ x 4-inch dowels, flush with lantern ends

1-inch center dowel extends 8 inches farther to support bearing and crank handle

1-inch center dowel extends into socket in base

⅜ x ¾-inch wooden dowels ½-inch inserted, tops rasped to taper radial surfaces

Figure A.8a & b. My particular demonstration crown-and-lantern gear pair. Not shown is a metal hook screwed down into the hub of the crown, used to pull the hub so the large gear can be removed for safer transport.

Not shown in the figures are an additional ¾-inch-thick disk with a 1-inch hole beneath the center of the crown and a 1½-inch block with a similar hole attached to the baseboard; these support and prevent wobble of the vertical shaft. Incidentally, while woodworkers five or six hundred years ago knew nothing of our fancy but cheap plywood, that modest modern material will provide a good measure of insurance against untimely splitting of the crown gear. (They did know of the advantage gained by joining flat pieces face-to-face with their grains at right angles—that goes back to very early wheelwrights.)

Chapter 5: Or Being Encircled

Archimedean screw pump. Perhaps the simplest way to make an Archimedean screw pump would be to insert a large (at least 1 inch in diameter) helically fluted wood-boring auger drill bit into a close-fitting tube, driving the bit with an electric drill. I tried just this and found that the arrangement worked very poorly—the usual pitch of drill helices is too great relative to the leakage between auger and tube. But tightly wrapping some flexible material around the auger so that auger and tube rotate together produces a functional pump. Cellulose acetate sheet serves well as the flexible material; rubber bands hold it to the auger, as in figure A.9. Since the pitch of the auger is on the high side for a pump, you can't use it with much inclination.

A 3- or 4-inch posthole digging auger (earth auger) makes a more impressive version, with not just a wider diameter but with a lower pitch—so it can sustain a steeper inclination. It can be fitted with an amply watertight seal by splitting a PVC pipe lengthwise, inserting the auger into half, and sealing it to the half pipe with silicone rubber sealant. Then liberally dope the remainder of the auger's edge and close with the other half of the pipe. Silicone rubber adheres poorly to PVC, so the assembly can be cracked open and the auger recovered if need be. But this version is more costly and can't be powered

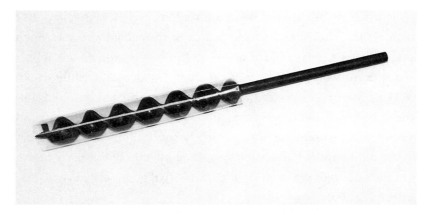

Figure A.9. A quick version of an Archimedean screw pump made from an old wood-boring auger bit that might be attached to an electric drill or other motor.

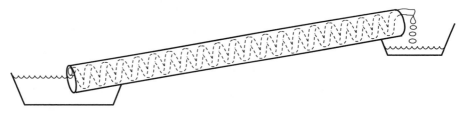

Figure A.10. Another version of an Archimedean screw pump. Note that the maximum slope is that which still allows the descending portions of the up-running hose to drain downward in the overall upward direction of the slope.

by ordinary electric drills—you'll need either a gasoline engine attached to some belts and pulleys or else some cranking arrangement.

A more effective, if less historically realistic, design (fig. A.10) will also prove the concept—I'm taking the idea from several websites that offer commercial models and more complex versions. Coiling a garden hose into a PVC drain/sewer pipe of perhaps 4-inch diameter provides an equivalent helical channel. This will work only with the thinnest, cheapest hose; better hose doesn't take so kindly to the degree of bending required. Something called "recoil air hose" or pre-coiled self-retracting garden hose does the job more easily; a 50-foot coiled length of ⅜-inch ID (internal diameter) slips nicely into a 5-foot length of that 4-inch PVC pipe used for the clepsydra (below).

The pump—if it is to be operated by turning the outside, treadwheel-fashion—requires some superstructure, which I leave it to the reader to improvise; this one is too small to tread in the literal sense. Cleats for turning can either be glued-on pieces of lengthwise-halved smaller PVC pipe or wooden strips held by long (or several joined) hose clamps. The flared (belled) end of the 4-inch pipe can be sliced up to yield collars that will keep the tube from slipping downward toward the lower basin. Finally, note that the screw has a preferred direction of turning to be effective, just like any common twist drill.

Treadmill. Treadmills are now standard fixtures of fitness centers and not uncommon in homes. The belts of the normal kind move under the impetus of an electric motor. If you walk or run on one, your power output increases as you increase the grade. So where does the metabolic energy you're expending go? Most, 80 percent or more, disappears down the sinkhole of your own energetic in-efficiency. You can calculate your power output by multiplying the vertical component of your speed by your weight—if speed is in meters per second (upward, remember) and weight is in kilograms times 9.8 meters per second squared, output will be in watts.

One might expect that some fraction of that power output might be picked up as a reduction in the current the treadmill's motor draws from the local electrical supply. I tested this by purchasing a clip-on ammeter ($15–30 and a nice thing to have around) and encircling one lead of a treadmill's power cord. As it happens, simply getting on the treadmill, at any speed or inclination, greatly increases the current it draws, almost certainly as a result of the increased friction between belt and underlying plate. As a result, any current reduction as the incline is increased is undetectable—at least as I attempted to measure it. Power (watts, again) for this purpose can be approximated as current (amps) times voltage (usually 120).

Another view of human work output takes advantage of an odd accident of unit conversion and contemporary standardization. Escalators in common use all seem to go at about the same speed and have steps of about the same height. If you maintain your position while walking up the down escalator, your power output in watts is approximately equal to your weight in pounds. Never mind metabolic rates—we're looking at power outputs not power consumption. So the exercise (literally) should not be undertaken by the unfit, the corpulent, or the aged. Checking my calculations for that coincidence ought to be done—at least for practice in unit conversion.

Chapter 6: Grabbing Again and Again

Grasp and release. Showing that one can turn a vertical grip in one direction more forcefully than the other takes only the simple device shown in figure A.11. The same resistance (rubber bands or spring) can be switched from one side to the other, avoiding the problem of matching them. Each subject should be tested both ways. A right-handed person should be better able to turn clockwise; a left-handed person counterclockwise.

Ballistae. Very small ballistae can be purchased in kit form, but it's not hard to make one's own, and the whole process is instructive. The finished product makes not just a good demonstration for people of any age (my four-year-old grandson was enamored with it), but the starting point for both discussion and specific calculation. What follows (fig. A.12) is adapted from material I contributed to the teaching website of the Society for Integrative and Comparative Biology, www.sicb.org/dl/.

For present purposes, one might replace the capstan shown with a more explicit system of sockets and handspikes. More specifically, I suggest a ring of holes in each end, with the holes in the ends out

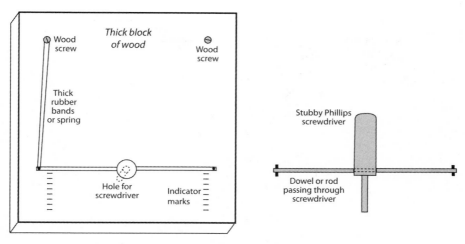

Figure A.11. A simple device to show that a person can turn a vertical grip more forcefully in one direction than in the other. The linearly elastic element (rubber bands or spring) insures direct proportionality between the force exerted and the indicator mark the rod reaches.

of phase. A pair of separate, long handspikes (whole-body levers if it hadn't been scaled down) would then be handled by the shooters.

The mode of energy storage puts very high stress on the front frame—elastic storage (roughly) follows the product of force and distance, and using an elastic material of minimal extensibility demands operation at high force levels. So the frame and tension-adjusting bars should be substantial. Do not lubricate the bearing surfaces between the rear frame and the revolving capstan. With luck, wood against wood will give sufficient friction so that a ratchet or other catch mechanism will be unnecessary. With less luck, one may have to drill a small hole down through one of the supports for the capstan for a rod that will stick into one of a circumferential ring of holes in the capstan.

Use braided rather than twisted rope to avoid any concern with asymmetrical behavior when twisted. I've used nylon rope; nylon (and most other common rope material) has a much lower resilience than tendon or any other good elastic. That degrades performance,

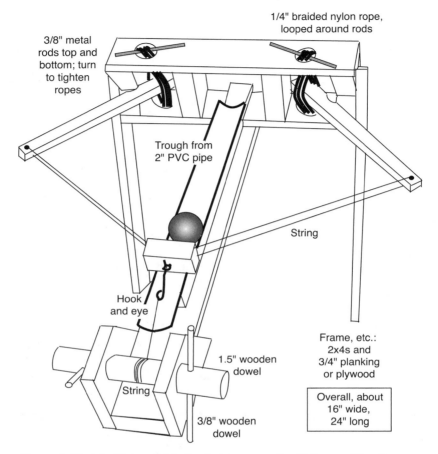

3/8" metal
rods top and
bottom; turn
to tighten
ropes

1/4" braided nylon rope,
looped around rods

Trough from
2" PVC pipe

String

Hook
and eye

1.5" wooden
dowel

String

3/8" wooden
dowel

Frame, etc.:
2x4s and
3/4" planking
or plywood

Overall, about
16" wide,
24" long

Figure A.12. A ballista suitable for desktop use—it will throw a Ping-Pong ball or perforated practice golf ball about 15 feet, beaning any sleeping student in the back row without fear of legal retribution. Pre-stress the ropes by twisting the metal rods; pulling the ballista arms back with the capstan will add the final store of energy. The hook and eye function as a trigger.

in present practice perhaps an advantage. It also allows one to ask about ways that the device could be made more efficient. A circular hole saw (available in any hardware store) on a drill or drill press simplifies making holes in the front frame for the ropes and in the rear frame for the capstan.

Note that the four rods at the tops and bottoms of the loops of rope are the only items that demand metallic material—as the ancients understood, those bundled tendons exert remarkably great force, even when not given the necessary extra pre-shooting twist. And the pieces that make up the front frame should be of substantial stiffness: ¾-inch plywood is just barely good enough.

Trebuchets. These may be purchased as desktop kits and, like ballistae, can also be easily built. As construction projects, they've enjoyed far more popularity, enough so that, rather than suggesting a specific design, I'd recommend a quick search of the Internet, but with the comment that a proof-of-concept model needn't involve the niceties of carpentry of most of what I saw. In any case, I find trebuchets of less mechanical interest than ballistae even if they certainly scale up more easily. For present purposes, the two make about the same point, so you might want to build only one or the other. In any case, the same hook-and-eye trigger and the same capstan and handspikes will work for both. Because of the descending weight, a trebuchet of equivalent capability will be heavier than a ballista, but since the weight descends and outweighs arm and projectile by a considerable factor, recoil will be similarly low.

Almost all published and posted designs lift the weight by means of a crank; like ballistae, serious medieval trebuchets are much more likely to have used handspikes or even treadmills. After all, these were devices that (as noted in chapter 6) lent themselves to— indeed, depended upon—whole body input.

Chapter 7: Turning and Unturning

Bow drill. A bow drill, as in figure 7.2, is a simple and forgiving thing to improvise. A long-shaft steel drill or a short drill with an extender will have a troublesome lack of friction when wound with the string of the bow. It's better to drill an axial hole slightly smaller than the drill in the end of an appropriate diameter (say, ½-inch)

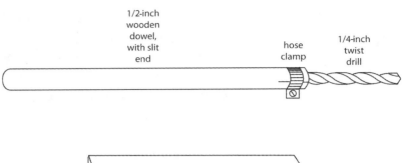

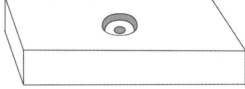

Figure A.13. (*Top*) A simple way to mount a modern drill bit in a wooden shaft suitable for a bow drill, pump drill, or top drill. (*Bottom*) An easy hand-piece for a bow drill or (as will appear below) a top drill.

wooden dowel, along with a lengthwise diametrical slit; a drill can be secured with a small hose clamp around the dowel (fig. A.13). Any bendable piece of wood or metal with holes near its ends will work as a bow; natural fiber cord does better as the bowstring (again because it gives more friction) than synthetics, although synthetics will work adequately.

In addition to the drill shaft, a hand piece is necessary. One easy and satisfactory grip that provides good support plus easy rotation of the shaft consists of a short length of 1 × 2 with a shallow ¾-inch hole drilled with a flat wooden bit. Into the hole one then glues a ¾-inch metal washer with a ⅜-inch central hole.

Spring-return grinding wheel. Springy saplings may not be commercial items, but we do have something our forebears lacked — nicely springy fiberglass fishing rods and (less commonly) tent stiffening poles. So one can make a spring-return horizontal grinder from contemporary components, as below (fig. A.14a). The main peculiarity will be obtaining or making bearings for the shaft. Conventional sleeve bearings can be purchased online for standard shaft sizes such as ½-inch diameter, or you can drill holes in hardwood and grease liberally. Further verisimilitude can be added with a foot-operated treadle. Passing the cord through a hole in the shaft that bears the grinding wheel will keep the cord from slipping uselessly around it; roughening the shaft may suffice. (Since the treadle will have a short travel and the pole a long bend, a form of Chinese windlass, as in figure 10.12, might be used on the shaft of the grinding wheel.)

In theory, this arrangement could drive a wood lathe, but I suspect that the requirements for adequate function would be a lot fussier.

Gravitational-return grinding wheel. For most of us, a counterweight return lends itself to improvisation better than does a spring return. The grinder just described works at least as well with a counterweight—I happened to use an old window sash weight, but a small sandbag would work as well. For grinding, only one free hand is really necessary, so one can dispense with the treadle, instead pulling a handle at the end of a rope with one's non-dominant hand, as in figure A.14b. (The arrangement is practical enough to attach to the end of a workbench as a regular fixture—especially a workbench with no nearby electrical supply.)

Pump drill. A pump drill (fig. A.15) is only a little trickier both to make and operate than a bow drill. It can use a wooden dowel bit holder similar to that of the bow drill, but now with a fitting such as a screw and pair of eyes (or some wire loops) at the top for the strings. Strings have to be the same length on both sides, and the hole for the crosspiece should be big enough so that the dowel fits quite loosely—otherwise any little asymmetry in pushing will cause the thing to lock up. One can either push inboard of the attachment points of the strings or extend the cross-board to allow an outboard push. A grinder wheel makes an excellent flywheel.

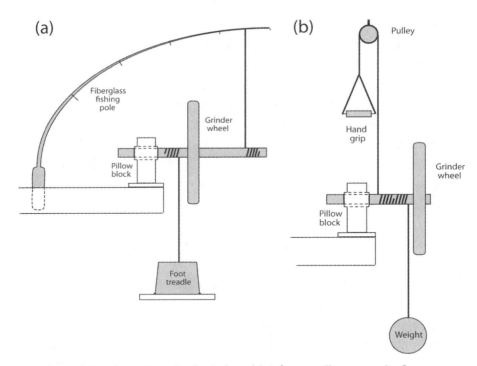

Figure A.14. Reversing-wheel grinders. (a) A foot treadle powers the first, with energy for the recovery rotation stored as flexion of a stout fishing pole. (b) A hand grip pulled downward powers the second, with a descending weight driving the recovery rotation.

Top drill. Once you have made a bow drill and a pump drill, rearrangement into the top drill described by J. D. McGuire (fig. A.16) is easy enough. This last takes the top support of the first and the flywheel of the second, adding only some screw, eye, or hole in the vertical shaft to which the string can be attached.

Chapter 8: The True Crank

Cranks. Few of these devices are simpler to make or more fun to try than Sleeswyk's hypothetical partially guided crank drill. Figure A.17

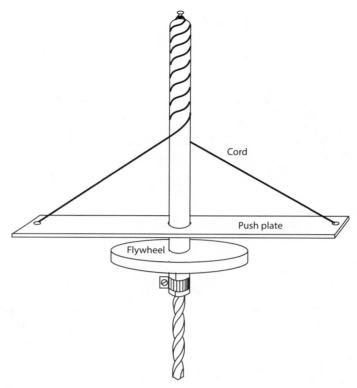

Figure A.15. One version of a pump drill.

gives construction suggestions. The same dowel, bit, and hose clamp can once again be used—or, as shown in figure A.17, you can use a conventional drill and a drill extension shaft; since this device operates at lower revolution rates, it will handle wider drills. A ½-inch aluminum bar works well for the crosspiece, although it could be made of wood or of cylindrical stock, and the drill extension can be fixed with a set screw, a cotter pin, or a long machine screw. The offset for the crank handle can be surprisingly small; in any case the specific value makes little difference except to the "feel" of operation. A weight of one or two pounds is handiest. Wobble of the shaft

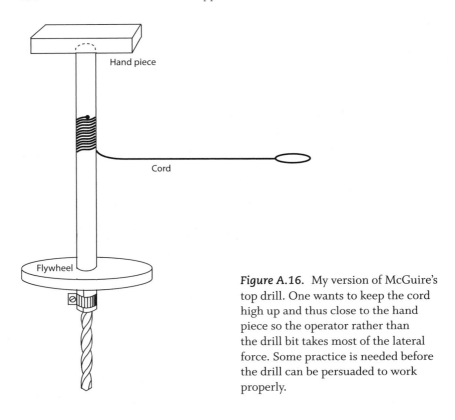

Figure A.16. My version of McGuire's top drill. One wants to keep the cord high up and thus close to the hand piece so the operator rather than the drill bit takes most of the lateral force. Some practice is needed before the drill can be persuaded to work properly.

can't be avoided, but it decreases with experience. If these things were, in fact, used as drills, they probably served only for shallow holes. (Don't be discouraged if operation seems impossible for the first few minutes—it takes an unusual motion, but almost everyone catches on after a little initial blundering.)

Chapter 9: Spinning Fibers

Shear stress and thread integrity. The ease of breaking a thread that comes from adding a drop of oil really has to be tried. Unfortunately, a more impressive demonstration happens to be unreliable—one can sometimes hang a weight from a thread just short of what's

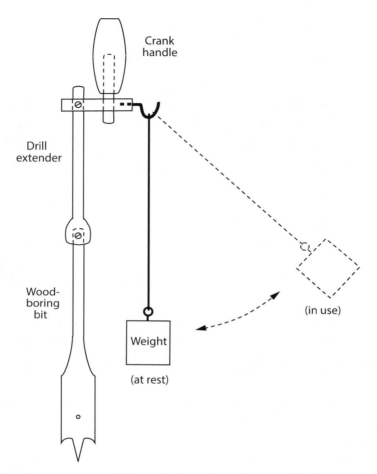

Figure A.17. A version of Sleeswyk's hypothetical Egyptian drill. Whether a correct interpretation or not, one of these devices works so much more effectively than one expects that it's worth making on that account alone.

needed to break the thread, touch the thread with a drop, and see it break. The difficulty comes from the non-uniformity of ordinary thread. A long length of thread and a long brushstroke with oil increases the odds of the thing working. Try about 500 grams or 1 pound (or maybe just a little more) as a weight. Another demonstration sometimes works—if you hang a similar near-threshold weight from a thread, the weight will unspin the thread, which at some point will then spontaneously break.

As mentioned in chapter 9, one has to make sure one is using pure natural fiber thread (typically cotton) with no polyester core or other addition of long, artificial fibers.

The basic reason why spinning per se matters was so central that I put the demonstration in the text, with figure 9.2. So only a reminder is needed at this point.

Whorl-and-spindle spinning. This kind of spinning takes practice, so don't expect too much in the way of speed or quality. Absorbent cotton or natural fiber batting provides the starting material; the same flywheel used for the various drills (if not too heavy) will once again serve, now as part of the spindle. Figure A.18 shows one arrangement for a dangling spindle that receives unspun fiber from a distaff held above and wraps spun fiber loosely around its shaft—which now must be smooth if the new thread is to slip off the end properly. Unlike wheel spinning, it's a start-and-stop process, with the spindle spun to twist the fiber above it, which is then rolled onto the spindle and run through the hook at the top while the spindle is again spun. A ½-inch dowel makes a good spindle. A large jar lid filled with mason's cement (see fig. A.6) is one way to imitate a stone whorl.

Chapter 10: A Few More Turns

Zero-angular-momentum turns. A turntable such as a large lazy susan will hold a not too heavy person. You (if you are that person) can crank yourself around by extending your arms, one forward and one back, turning them both clockwise, pulling them in against the body, turning them counterclockwise, extending them again, once again turning them, extended, clockwise, and so forth

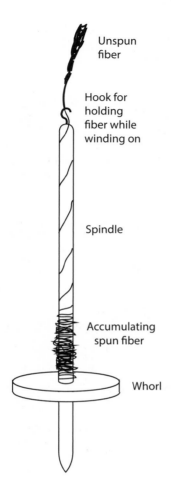

Unspun fiber

Hook for holding fiber while winding on

Spindle

Accumulating spun fiber

Whorl

Figure A.18. A drop-spindle hand-operated whorl-and-spindle spinning rig. A distaff isn't shown; it would be held above in the other hand. Distaffs came in even less stereotyped forms and could be little more than a loose bundle of carded fiber.

(see fig. 10.1). Holding weights in the hands—even 1-pound cans—improves matters considerably. Another way to execute a zero-angular-momentum turn consists of clasping both hands around a weight, perhaps two or four pounds, and turning hands and weight in a horizontal circle, waist high, in front of you. (Doing a zero-angular-momentum turn while seated in a swivel chair seems to work less well because the system has much more rotational inertia

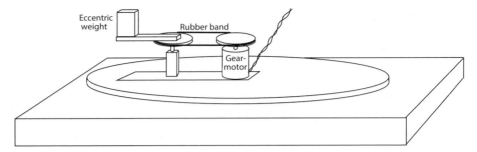

Figure A.19. A device to demonstrate zero-angular-momentum turns that uses a surplus record player turntable as a low-friction base. See text for details.

than a person standing erect. Thus you change it proportionately less by your movements.)

You don't really need a person to do the trick. If you have an old turntable, you have a nice stable rotor with a good thrust bearing. If you can, disconnect any circumferential drive belt from its driving motor, and certainly don't plug it in. You then need to construct an eccentric appliance that sits on top of the turntable, as in figure A.19.

Most convenient is a small gearmotor of the kind used in vending machines and similar applications—a motor turning at about a revolution per second is about right—together with pulleys normally used for sliding doors. Since most gearmotors are fairly heavy, it should be centered on the turntable. From the gearmotor, some belt drive or other radial extension needs to turn the vertical shaft on which the off-center weight will rotate as a sideways-extending shaft with some mass at its far end. Its rotation should have the about the same radius as the radial extension so that, in each revolution, the extra mass passes across (or near to) the turntable's center. A $9/32$-inch or $5/16$-inch hole in the plate that supports the motor and pulley that engages the spindle of the turntable will position the appliance. The power cord for the motor might conveniently dangle from above, meeting the motor opposite the eccentric shaft. When turning, the action of the gearmotor should crank the turntable around with the kind of jerky motion you experienced when moving weights yourself. Since the turntable will not rotate with any great speed, you needn't worry about the power cord twisting too much or bother arranging battery power instead.

Irrotational vortex. A circular plastic tub draining through a hole in a standpipe near the middle of the tub will produce an approximation of an irrotational vortex, as in figure A.20. How can you show that the vortex is irrotational? Float something that will not rotate on its own in the tub as the tub empties; after emptying has proceeded long enough for the vortex to develop, the float should translate around the drain, maintaining its heading as it circles. By contrast, if the hole is kept plugged and you stir the water to make it go around (a rotational vortex), your float will rotate as it circles.

In practice, the demonstration turns out to be a bit fussy. I equipped a 17-inch plastic tub with ½-inch plastic pipe fittings (nominal size; close to ¾-inch actual outer diameter), with a female socket at the inside bottom from which a 6-inch nipple could be unscrewed

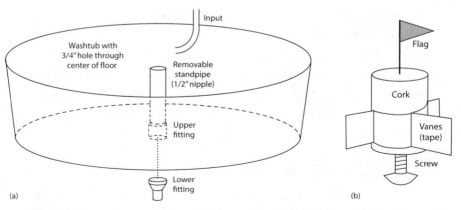

Figure A.20. Irrotational vortex generator. (a) The tub and its fittings; (b) a float that will show the presence or absence of local rotation.

to start the draining (fig. A.20a). The whole thing should be positioned (perhaps on bricks in a large sink) so the protruding center bung isn't impeded. A float can be made from a short wine-bottle cork with a screw in its bottom heavy enough so most of the cork will stay submerged (fig. A.20b). Vanes on the float prevent self-rotation, and a flag on top gives its orientation at a glance. Besides the float (or if it proves uncooperative), sprinkling sawdust on the water's surface can show the swirl of the irrotational (or near-irrotational) vortex. A few drops of food coloring added at some point may be instructive as well. I initially filled the tub about 4 inches deep.

Keeping a steady and well-directed input flow at the periphery while water drains often improves matters. The input should enter through a wide diameter opening and thus at a low velocity. Initially, the hole should be plugged and the tub filled with water. The residual swirl from filling ought to be sufficient to establish the direction of the vortex; if not, a slight initial stir will do the job. (Never mind the Coriolis effect; showing the hemispherically dependent bathtub vortex trick takes heroic effort.[1])

Weissenberg effect/strangulated flow. Wiegand (1963) gives instructions for what amounts to an overly elaborate demonstration; YouTube videos suggest easier alternatives. I find that I can achieve modest strangulation with Purell hand sanitizer (liquid, not foam), using a Dremel tool with a cylinder of ⅛-inch diameter in the chuck turning at less than the tool's maximum speed. Corn syrup has about the same gloppiness, but it doesn't strangulate and thus provides a nice comparison. More impressive (and less fussy) strangulation comes from a thick slurry of methyl cellulose (I used Dow Methocel, as in fig. 10.3).

Twisted-rope vise. The remarkable force that a twisted-rope vise can exert can be demonstrated with the simple apparatus in figure A.21. The main safety caveats are (1) avoid using any stretchy

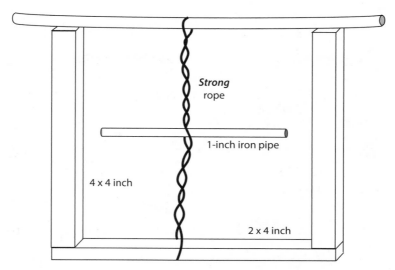

Figure A.21. A twist-rope vise—demonstration version. Again, be careful with one of these. I once broke some rope that I had thought more than amply strong (with quite audible release of stored energy), rather than fracturing the intended sacrificial wooden dowel.

rope so the energy accumulated in the rope as it develops tension is minimized; (2) break something that, again, stores little energy, such as pressure-treated wood rather than untreated wood; and (3) wear safety goggles when twisting the rope. Wooden tomato poles, incidentally, are a good source of pressure-treated wood of a convenient size. (Pressure-treated wood breaks more cleanly and with less accumulated strain energy.) Equipped with a strong tension gauge, the device could be used as an inexpensive tensile-testing machine for loads too great for hanging weights to be handy.

Bucksaw. The bucksaw of figure 10.11a might be put to practical use around your yard if equipped with a tree-trimming blade. Or a small version makes a perfectly fine hacksaw—it's not just as good an illustration of a rope-twist vise. Figure A.22 gives details for one version

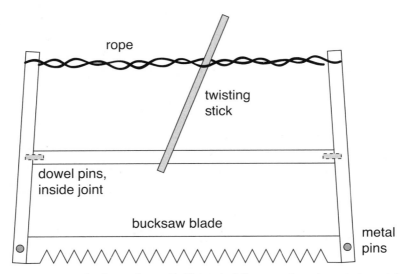

Figure A.22. A bucksaw that will obligingly fall apart when the twisting stick is removed, showing that all parts are secured by nothing more than their interacting and properly balancing forces.

of the first. Note that it needs no fasteners, and the two pins that take the concentrated shear force at the blade's end are the only metal elements other than the blade. (Long screws will serve for these pins but function simply as shafts.) Note also that this device illustrates all the ordinary modes of loading. Thus:

> The central crosspiece takes compression.
> The blade at the bottom takes tension.
> The short metal pins connecting blade and side pieces take shear.
> The twisting stick is subject to bending (flexion).
> The rope at the top is torsionally loaded.

In practice I used 1-inch-square pieces of hardwood for the crosspiece and sides, ⅜-inch dowel pins of wood to keep the crosspiece positioned, and ¼-inch-diameter steel screws threaded only near their ends to tie blade to frame. The twisting stick was a piece of ½-inch dowel.

Differential windlass. Making a differential or Chinese windlass (fig. A.23) presents few problems; as with the twist-rope vise, operation drives home the reality of just how much force can be produced with a very simple device. I used two cement-filled grocery store cans, an 8-ounce tomato sauce can, and a 6-ounce tomato paste can to produce the different diameters of shaft needed. Filling the cans with cement was done with the dowel and cans vertical and the long

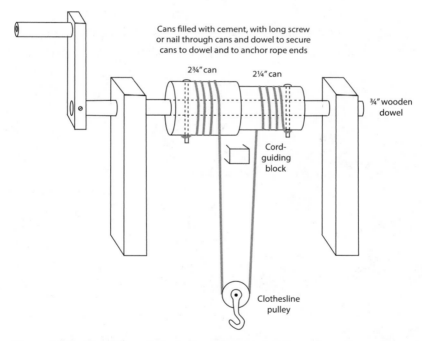

Figure A.23. The differential windlass. I've not shown any lower mounting; it might be no more than two adjacent tables or a pair of sawhorses. The cord-guiding block (attaching to the lower mounting) is an ad hoc device to keep the cords winding on their proper drums.

screws already installed; a little caulk around the bottom dowel holes minimized dripping before the cement hardened.

Chapter 11: Rolling Back Rotation

Steam-jet engine. A pure steam-jet engine, with the same basic drive Hero used but with no rotation, was once trivial to make from the small metal containers that held rolls of 35-millimeter film. Then the containers became plastic, and finally the film itself largely disappeared. Even better, though, are cans from the small-size Sterno jellied alcohol, available in some supermarkets and hardware stores. The contents are water soluble and clean out easily. The same product and size serves as fuel. And the press-fit tops provide a measure of safety—a top will blow off if the pressure inside becomes too great—so be sure *not* to permanently affix the lid. Figure A.24 gives a possible setup. (One can drill a tiny hole [about $\frac{1}{16}$ inch] in the top

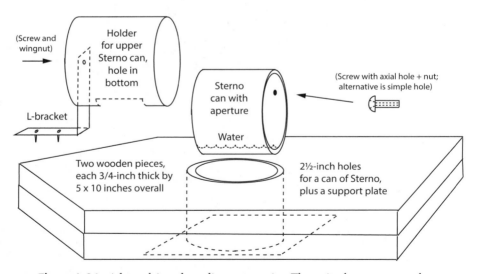

Figure A.24. A boat driven by a direct steam jet. The point here reverses that of almost all the previous models in this section—this boat works but does so very badly.

of the can for the jet, but a little better is a hole through the axis of a short screw, with a taper [from reamer, countersink, or other drill] on the input side.) A metal can with the lid removed, with a diameter slightly larger than that of the Sterno can, makes a convenient receptacle for the latter—an 8-ounce tomato sauce can, 2½ inches in diameter, fits perfectly. Put a couple of teaspoons of water in the "drive" Sterno can as a source for the steam before inserting the Sterno can in the holder.

What's of interest here is just how poorly the boat performs— try it in a bathtub or some indoor body of water, since even small amounts of wind will overwhelm the jet. Yes, the Hero engine works, but its propulsion efficiency could scarcely be lower.

After use, slide the top onto the container of fuel and let the top can cool slowly. If you throw water on it, it may collapse from the unbalanced atmospheric pressure on the outside.

Rocker-arm motor. A version of Joseph Henry's rocker-arm motor made with contemporary material (fig. A.25) presents few difficulties. A pair of 9-volt batteries will supply power; like his original wet cells (figure 11.5), they have both terminals on top. Flat magnets polarized top to bottom are handiest; these are common, inexpensive bits of hardware used to temporarily attach such things as spotlights to metal doors. (Those in the drawing are ceramic block magnets from Harbor Freight Tools, item 97504, but they hold no special attraction beyond their low price.) For the pivot, a transverse nail or screw and a V-groove in the mounting should be sufficient. Appropriate electromagnets to attach to the beam can be made by winding about 400 turns of 30-gauge enameled copper wire (a Google search will reveal a number of sources) onto plastic sewing machine bobbins. (Don't forget to scrape the insulation off the free ends of the wires.) The polarity of the battery-electromagnet pairs and their respective magnets should be picked to ensure that contact causes repulsion. You can turn the motor on by lowering the beam onto the pivot or turning a battery in its clip.

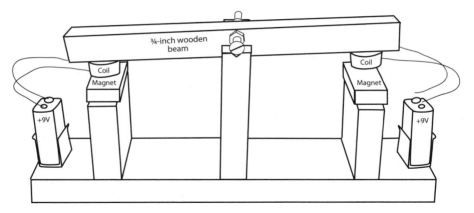

Figure A.25. Building Joseph Henry's motor with off-the-shelf material. The only part that's at all ticklish is getting the ends of the flexible wires to make contact with the battery terminals at about the right point for the magnets to give proper pushes.

Hydraulic ram. Figure 11.8 gives a diagram of a hydraulic ram. For instructions on building one with off-the-shelf components, I can't do better than what you'll find at http://www.clemson.edu/irrig /equip/ram.htm.

Clepsydra. A simple clepsydra is given in figure A.26. By making the rate-limiting orifice operate entirely underwater, one eliminates any influence of surface tension, a potential source of trouble. The ancients had to contend with the severe dependence of viscosity on temperature—flow is proportional to viscosity, which drops as temperature rises. Living in thermostatically controlled homes, we should find clepsydrae more reliable—if still far worse than escapement-regulated clocks, to say nothing of electric and electronic ones. The intrinsic non-linearity caused by the decreasing net weight (on account of buoyancy) as one part descended seems to have given fits to the ancients—a taper that offsets that non-linearity isn't an ordinary one. Modern materials can minimize that problem by concentrating the weight so most of it is either always or never submerged. What's offered in figure A.26, then, is a clepsydra that models no ancient design but might well time, say, a lecture on clepsydrae. I've omitted some obviously necessary supports; these will depend on where one wants to place the device and whether it's intended to be movable. Note (before starting) that the thing holds about 20 pounds of water and stands well over 6 feet high. I hold the one I built down to its wooden platform with four L-brackets screwed into the platform; a large hose clamp encircling the tops of the L-brackets and the base of the clepsydra ensures stability.

The smaller the pore (offset by a heavier weight), the more nearly

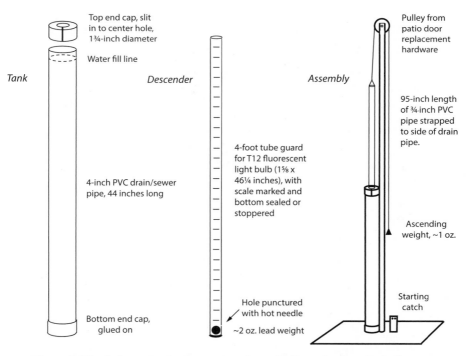

Figure A.26. A clepsydra for the modern household; notice how little floor space it occupies. A little ingenuity will ensure that the ascending weight triggers some announcement that it has arrived at a predetermined terminal point—for instance, a knot in the string might fail to rise through a hole in a hinged plate on a switch, flipping it.

linear will be the performance of the device. One resets the clepsydra by lifting the descender so its water will pour back into the larger cylinder, adding a little if needed to make up for evaporation. After releasing the weight from the starting catch, the descender will drop to an initial, zero-time position. Removing the water to move or store the device goes easiest with a pump connected to a sufficient length of flexible hose on its input.

[Notes]

Chapter One

1. Full et al. (1993).

2. Having a minimal head and no heels at all. Brackenbury (1999).

3. The view, not the test of one's esophageal sphincter, was the attraction. George Washington Gale Ferris Jr.'s 1893 original, 264 feet at the top, took nine minutes for a revolution.

4. James (1906).

5. Ramelli (1588). This monumental compendium, available in an excellent translation, will supply us with examples at many points in chapters to come. It's worth looking at simply as a fine example of the logic of an advanced wood-based rather than metal-based technology.

6. And might I put in an admonition about helping to fund *Wikipedia*? It's all too easy to take for granted this spectacular resource.

7. Johnson (1992); Steele (1994).

8. Smeaton (1759).

9. Lilienthal (1889).

10. Anderson (1997).

11. Specifically, by George Santayana (1863–1952), "Those who cannot remember the past are condemned to repeat it."

12. Berg and Anderson (1973).

Chapter Two

1. To be specific, Eratosthenes' long life seems to have encompassed most of the third century BCE—according to a fine book on him and the context of his work by Nicastro (2008).

2. *Wikipedia* has a good article; I do take exception to the high efficiencies cited for propulsion by oscillating appendages: en.wikipedia.org/wiki/Rotating _locomotion_in_living_systems.

3. Full et al. (1993).

4. See www.youtube.com/watch?v=biuoagnGCFQ; also, while at it, see www .youtube.com/watch?v=HmLS2WXZQxU.

5. Henschel (1990).

6. The trees whose wood bears the name are indigenous to the Caribbean and adjacent South America. The name "lignum vitae" alludes to the supposed medicinal value of the resin, the impetus for its import into Europe.

7. Roosevelt (1914).

8. All this from McShane and Tarr (2007).

9. This comes from Lay (1992). It's a splendid account, admirably well documented.

10. Thurston (1894). Quite startling to me is how much good biomechanics and nutritional and respiratory physiology was known and appreciated, at least to that author.

11. A subtle point here is too easily misunderstood—as it was in a middle school science textbook to which my son was subjected. If you move a 100-pound desk 10 feet across a room, you have not necessarily done 1,000 foot-pounds of work—the actual work depends on the friction with the floor; and with good wheels on the desk might approach zero. What matters isn't the weight of the desk but the force opposing your effort to move it—friction, gravity, whatever. (You do have to get the desk moving, which means accelerating its mass; and that does take some work.)

12. Wilson (2004).

13. Lay (1992); Piggott (1983).

14. Anthony (2007). The location has been questioned, with the Anatolian plateau (modern Turkey) suggested as a likely alternative.

15. Lay (1992). Since the values are relative to some standard, they are unitless—force and weight have the same units, whether pounds or newtons.

16. Taylor et al. (1972).

17. The data come from Baker (1903).

18. Diamond (2005).

19. Ekholm (1946).

20. For the origin and early history of wheelbarrows, see Lewis (1994).

21. Cotterell and Kamminga (1990). Again, this is the tractive force of the person, not the load in the wheelbarrow or other vehicle.

22. The ones shown in the great treatise on mining of Georgius Agricola of 1556 look particularly lift-demanding; at least those very forward wheels were fairly large.

23. Temple (1998). The underlying data come from Joseph Needham's great multivolume treatise, *Science and Civilisation in China*.

24. I give a more complete account in Vogel (2001).

25. White (1962)—among other of his accessible (and important and enjoyable) writings.

26. I especially like McShane and Tarr (2007). Enormous acreage was devoted to raising feed, removal of manure challenged municipal services, and the insects and odors were an ever-present reminder of who really ran the cities.

27. See http://qdrum.co.za/about-q-drum.

28. From White (1984), p. 79. A less radical but more practical solution put wooden circular frames—wheels, in effect—around the ends of the beam, linked by a circumferential frame with stub axles at each end.

Chapter Three

1. During 1943 the British and Americans expended a disproportionate fraction of their heavy bombing campaign trying to destroy the ball bearing factories at Schweinfurt, Germany—so vital were bearings to military hardware and its production. French (1988) gives a quick look at bearings, including those within our bodies.

2. Bear this in mind when you shop. Drawers in stores, naturally, lack loads, so they all slide smoothly and require little force to operate. Caveat emptor.

3. Oberg et al. (1984).

4. The story is given, with a host of other horological information, by Landes (2000). A highly readable account (but with all too few technical details) of Harrison's work, even if the title may exaggerate, is provided by Sobel (1995). Models of the chronometers can be seen at the Royal Observatory, Greenwich, in London.

5. Getting motion started, working against "static friction," takes more force than keeping it going, quite beyond the force needed simply to accelerate the body—once you start to skid or slide, you tend to keep going on account of the lower sliding friction. A good introductory source on the parent subject is Bowden and Tabor (1973).

6. Gebeshuber et al. (2008).

7. McLaren and Tabor (1961).

8. The name "lignum vitae" refers to its original medicinal uses, too numerous to enumerate and none of demonstrated efficacy.

9. A short introduction can be found in French (1988).

10. Oberg et al. (1984).

11. Or rotational kinetic energy, the variable most often met in the literature on throwing pots on a wheel. While the two are not the same, the difference is of little importance here.

12. See, for instance, Childe (1954) or Choleva (2012).

13. Orton et al. (1993).

14. See, for instance, Foster (1959).

15. Childe (1954).

16. Wu et al. (2012) offer good evidence of pottery in China 20,000 years ago, long before reliable evidence of ceramic vessels associated with agriculture.

17. Kerr and Wood (2004). This continues, as volume 5, part 12, the comprehensive work begun by the great Sinologist Joseph Needham (1900–1995), now under the auspices of the Needham Research Institute, in Cambridge, UK. On the eccentric and fascinating Needham himself (but with all too little consideration of his monumental work), see Winchester (2008).

18. The rule, known as Laplace's law or the Laplace-Young relationship, says that the pressure difference a wall of a given material can withstand varies inversely with its radius of curvature—skinnier is stronger.

Chapter Four

1. For just one source, see Kraybill (1978).

2. Carter (1980) is particularly useful on terminology.

3. Takaoglu (2005).

4. Caulfield (1977).

5. Hurt (1982) gives some illustrations, as do issues of old farm equipment catalogs and publications such as *Prairie Farmer*.

6. Van Bueren (2004).

7. Thurston (1894); available from http://kmoddl.library.cornell.edu/bib.php?m=53.

8. Oberg et al. (1984).

9. Smith (2006) provides a wonderful analysis of the complex mechanical and organizational aspects of pyramid construction, bringing to the issues his own background as the managing engineer for large-scale public works projects.

10. Agricola (1556).

11. Ibid., book VI, p. 164. The Hoovers have thoughtfully converted his measurements into more modern units.

12. Ramelli (1588). The translation is excellent, and one marvels at the quality of the draftsmanship of the figures—well beyond merely leaving no ambiguity about the mechanical functioning of the devices. These approach the level of the plates of Vesalius's *Anatomy*, produced by the associates of the great painter Titian.

13. Ibid., plate 138.

14. Prager and Scaglia (1972).

15. Moon (2007) points out (with a figure) that one motorized toothbrush, the Colgate Crest "Spin Brush Pro" uses a crown wheel and pinion, the latter functionally the same as a lantern. Of course, it reverses the sequence, with the pinion as driver, to decrease rotation rate—small electric motors turn rapidly with little torque and more often need gearing down than up.

16. A swell resource both on the web and to visit is the American Wind Power Center and Museum, in Lubbock, Texas (www.windmill.com).

17. From http://kmoddl.library.cornell.edu. Another entry point is www .archive.org. The parent source for these sites was Moon (2007).

18. "The Archimedes Project" at http://dmd.mpiwg-berlin.mpg.de/home.

19. Ramelli (1588), chap. 174.

20. Major (1990) gives a good summary of these later developments.

21. A particularly readable account of the enterprise is given by King (2000).

22. Almost everyone seems to cite Stuart (1829) as the basic source, so all turns on his reliability, of which I'm no judge. See, for instance, Crisman and Cohn (1998) and Erickson (2006).

23. So much for the intrinsic superiority of biomimetic technologies!

24. Crisman and Cohn (1998).

Chapter Five

1. If you're using a whim as an exercise machine, you'll need some kind of adjustable brake to absorb your work and you'll face the same problem as did the ancients—speeding up the rotation rate of the braking device relative to that of the whim. One solution would consist of finding an old (and large) TV antenna rotator for the arm, using its motor as a generator, and adjusting the resistance connected in place of the motor's original power source.

2. Marcus Vitruvius Pollio (c. 80–70 BCE–c. 15 BCE), especially *De Architectura libri decem*.

3. Landels (2000) notes evidence that oxen or donkeys were used, presumably via whims, crown-and-lantern gears, and vertical bucket wheels, to lift water in second-century CE Egypt.

4. The Thorkild Schiöler collection (http://kattler.dk/schiolers/uk/) and Hodges (1970), fig. 211, p. 217. I find the former more persuasive than the latter, but they do look as if they illustrate the same thing. Landels (2000) suggests that the low tilt of the cylinder might be indicative of a mill or crusher rather than a water pump. But I think that anything other than a pump would have needed a higher-force, lower-speed driving arrangement than a person treading on top.

5. Vitruvius, Book X, describes these and the pumps that follow—a screw pump, the two drums, and the bucket chain.

6. Landels (2000).

7. Agricola (1556), book VI, p. 211.

8. Strada (1617–18), plates 78 and 94; the first powers a piston pump, the second (shown here) powered a gristmill.

9. Veranzio (1615), plates 23, 24, 41.

10. Ball (2006) puts the contemporary chemotherapeutic scene in fine context.

11. Ramelli (1588), plate 123.

12. Minetti (1995).

13. *Oxford English Dictionary*, 2nd ed.

14. I've given more elaborate accounts (if still fairly short descriptions) in Vogel (2001) and (2002). Further material (again short) can be found in Cotterell and Kamminga (1990).

15. Crisman and Cohn (1998) give the question more attention.

16. I recall, in particular, the one in the Harvard Square station of the old Boston MTA, before modernization and extension of the Red Line.

17. All this from a short paper by Tarr (1999), one with many good references.

18. Grant (1885), vol. 1. The river is the Ohio.

19. McGrail (2001), citing Peng (1988). The great source on such things is the multivolume, magisterial *Science and Civilisation in China*, begun by Joseph Needham—but on this matter it adds little.

20. For instance, Roberto Valturio (1472), an Italian military engineer. The reference is from Stuart (1829). Both are available in digitized form. The unlikeliness of reality for such a craft has already been pronounced.

21. Mark Twain, in *Innocents Abroad* (1869), traveled on one; his comments give some idea of the motions of such ships in heavy seas and their profligate use of coal. But the very occurrence of a pleasure voyage on one says something of their at least minimally satisfactory nature.

Chapter Six

1. At one time a large child's top made use of the same telescoping helix to set it turning. One just pushed down on the knob above the center, and its helically twisted metal shaft moved down into the body of the top through a fixed slot.

2. I strongly recommend that a bench (of appropriate height) with a vise be provided for youngsters by at least age five. Without a vise, their attempts to use tools will too often frustrate and discourage them; with one they'll get a feeling of reward and empowerment. Furthermore, from a fairly early age, they will be able to work with many adult tools, simply by using two hands where an adult would need only one.

3. Among other applications, a bomb calorimeter can be used to determine the energy content (calories or kilocalories) of food. The version that appeared in my physical chemistry class in 1959 was instructively basic—websites offer many fancy contemporary models.

4. McGrail (2001).

5. Ensor (2004).

6. We non-affluent had what we called "Armstrong Power Steering."

7. Agricola (1556), book VI, pp. 162, 171, 182, 194.

8. Ibid., p. 162.

9. Hacker (1997) summarizes a century of effort.

10. Tarver (1995) both analyzes the history and describes construction and testing of a full-size reconstruction of a traction trebuchet.

11. The Roman name, now the common term (where the generic "catapult" isn't applied); to a Greek, it would have been a "palintonon."

12. Landels (2000); Gordon (1978).

13. Marsden's (1971) translation of Heron's artillery manual of about 100 CE, p. 27, notes that the whole device is elevated off the ground, with holes in the capstan for handspikes. But Heron doesn't give the size of the particular ballista and implies that it isn't especially large. See also Marsden (1969).

14. Marsden (1971), p. 205.

Chapter Seven

1. Childe (1954). More recent sources don't disagree.

2. They can form the main elements of attractive mobiles; they're ordinarily discarded in large quantities since we serve only their adductor muscles for culinary purposes.

3. Data from Vogel (2013).

4. Coppa et al. (2006).

5. See Hurt (1982) and antiquefarmtools.info/page3.htm.

6. Charles Pell, who has a wonderfully sharp eye for all things mechanical.

7. Back in the early 1960s, seat belts were consumer add-ons. Automobile manufacturers, in their favorite foot-dragging fashion, deigned only to pre-drill the floor pans of cars to ease installation. Then retractors were add-ons to the add-ons. At least a person developed a good sense of how they worked.

8. See, for instance, Childe (1954), Woodbury (1961; 1972), and Rybczynski (2000). One problem in determining lathe origins is the difficulty of distinguishing between wooden objects cut while turning on a lathe and those rounded by turning slowly around a template and cutting (or filing or sanding) off the high spots.

9. The following site gives an audio report: http://minnesota.publicradio.org/display/web/2009/09/03/fair-sound-day6-info. For a more general view, see http://www.historicgames.com/lathes/springpole.html.

10. The best source I've found on the matter is Woodbury (1961). The great heroes are, in the United States, David Wilkinson, and, in the UK, Henry Maudslay, easily as important, if much less well known than, say, Fulton and Watt.

11. I recommend the arrangement as often preferable to the simple springs used for screen doors. The pull of the descending weight is almost steady (acceleration proves almost trivial), and opening the door (or whatever raises the weight) doesn't require ever-increasing force.

12. See, for instance, www.nationalyoyo.org, as well as the article about yo-yos in *Wikipedia*. Bürger (1984) introduces the underlying physics. He puts things in equivalent terms of translational and rotational kinetic energies rather than linear and angular momenta.

13. An example of the operation of a pump drill, at this writing: http://www.instructables.com/id/Primitive-fire-starting-using-a-pump-drill/.

14. See, for instance, McGuire (1896), Davidson (1947), and the online

collections of museums such as the Museum of Anthropology of the University of British Columbia.

15. McGuire (1896), p. 718.

Chapter Eight

1. In case you're interested: a pencil sharpener, an eggbeater, cranked-rechargeable radio and flashlight, food grinder, nut chopper, and ice-cream freezer—none at all existentially vital in their hand-cranked manifestations.

2. Landels (2000), pp. 10–11.

3. Childe (1954), pp. 192–93. But see, for an alternative view, the text and references in Lucas and Harris (1962), pp. 423–24.

4. Temple (1998), p. 46.

5. In the *Utrecht Psalter*; see Dewald (1932).

6. Prager and Scaglia (1972), figs. 25 and 89.

7. Ramelli (1588), plates 162–64.

8. Agricola (1556), p. 162.

9. Besson (1569–78), p. 150.

10. The sound of a hurdy-gurdy can be heard in any of several video clips that turn up in a search starting with the instrument's name.

11. The remains of the *Hunley* were salvaged in 2000; they can be viewed at the Warren Lasch Conservation Center, in North Charleston. In addition, the state historical museum in Columbia has a fine full-scale model.

12. A variety of audio clips are cited in the *Wikipedia* article "Glass harmonica." A good general reference is Zeitler (2013).

13. Not to be confused with Isaac Bashevis Singer (1904–1991), a very fine Yiddish-American author.

14. Lessing (1998), cited in *Wikipedia* article, "History of the bicycle."

15. A good source for the bicycle's early history, with a fine collection of contemporary illustrations, is Oliver and Berkebile (1974), which may be available for free downloading. Another good source is Herlihy (2004), who disparages claims of the reality of the usually noted MacMillan machine.

16. Figures for the maximum power output of muscle run from about 20 to 200 watts per kilogram, with figures from work on people at the low end. The similarity of muscle structure suggests that the low figure merely reflects unavoidable experimental limitations for work on human material. Thus I've used the higher data.

17. There's obviously more to the story. For a little more, see Vogel (2013); for quite a lot more, well explained, see MacIntosh et al. (2005); on bicycles in particular, see Wilson (2004).

18. Sleeswyk (1981). A search (Google Scholar) starting with his name returns quite a few papers that provide interesting physical and engineering interpretations of items of ancient technology.

19. Under "Pitman arm" in *Wikipedia* one learns about a key component of a car's steering system, one that's truly a pitman arm; under "sawmill" one learns

about "pitman arms." The eponymic Pitman, capitalized, attaches to a system of shorthand writing devised by Sir Isaac Pitman.

20. An excellent working model can be observed at Old Sturbridge Village, Sturbridge, Massachusetts.

21. Ritti et al. (2007).

22. An animation can be seen in the *Wikipedia* article "Boulton and Watt steam engine (Powerhouse Museum)"—or, if you're in Sydney, Australia, see the actual engine, fully restored.

23. Bressel and Heise (2004).

Chapter Nine

1. Ryder (1968). Forbes (1964) talks about interlinking surface irregularities, as does Crowfoot (1954) in almost the same words.

2. The demonstration is a little less effective with a knitted washcloth.

3. Very long leather straps and laces were made by cutting hides spirally, so the length of the tanned hide did not limit their lengths.

4. Make sure you're pulling on a thread of 100 percent cotton—in particular, avoid silk or synthetics such as polyester—their fibers are very long and thus threads and cords made of them gain little beyond physical coherence from being spun.

5. But if you do hit the fabric store for starting material, again don't buy synthetics.

6. I'm struck by the wide-ranging speculative character of the anthropological literature and by the relative absence of sharp partisanship. See, for instance, Crowfoot (1954), Ryder (1968), or Barber (1991).

7. Gandhi, one of the true greats of the twentieth century, was no economist. Clothing a country with homespun, as he advocated, would have been so labor-intensive as to have imposed a fair level of national poverty. We should not be surprised that the large-scale, powered spinning in factories was an early hallmark of the industrial revolution.

8. Good descriptions of the process are given by Lemon (1968) and Tiedemann and Jakes (2006); Patterson (1954) offers a shorter but nicely comparative one.

9. Quoted in Lawler (2014).

10. What one sees demonstrated in museums and reconstructions are these highly evolved wheels; one should not be surprised that their modi operandi appear less than obvious.

11. White (1974). White devoted much of his long career to rescuing the medieval period from its reputation as the "dark ages," documenting that its rate of technological changes was greater than any preceding period. On the matter of spinning wheels, see also Marchetti (1979).

12. While our word "paper" comes from "papyrus," the material papyrus had serious drawbacks as a medium for accepting and preserving writing. It wasn't cheap to produce from the papyrus plant, and the product lacked both the flexibility and durability necessary for widespread dissemination of written material.

Paper made exclusively from wood pulp—*really* cheap paper—is a nineteenth-century achievement.

13. As Robert Temple (1998) points out, large-scale paper production began much earlier in China; mass printing followed shortly thereafter. The similarity of sequence increases confidence in a facilitative relationship.

14. The numbers have been collected and reduced to relative homogeneity by Tiedemann and Jakes (2006).

15. The references to asbestos cloth in classical texts have been largely ignored in recent times. But an old account, Gilroy (1853), quotes many sources.

16. See *Wikipedia* articles "Silk production" and "Sericulture."

17. A few other saturniid moths provide commercial quantities of silk, most notably the tussah silk moth, *Antheraea perni*. Tussah or "wild" silk can't be unwound; instead, chopped-up cocoons yield fiber that must be spun in the same way as cotton or wool. *A. perni* can survive in the wild; *Bombyx* is as much a product of human breeding and domestication as is modern corn.

18. Kirby and Spence (1815–26). The reference to the spinneret is in vol. 3.

19. LaBarbera (1985). He has experimentally worked out what may have been the way they did it.

20. Gordon (1978) gives a good mechanical explanation; McGrail (2001) puts the Egyptian vessels in a historical account of boats.

21. Gilbert (1954).

22. "The Ropewalk," quoted as front matter in *The Story of Rope*, published by the Plymouth Cordage Company (1916). The datum on number of ropewalks comes from the same source, quite a fine (and free) volume.

23. I don't want to expend the considerable space necessary to properly explain the way the ropewalks operated. A fairly good video of the one at Chatham Historic Dockyard is given at http://www.youtube.com/watch?v=2M5mo2I2c0Q. The best simple description I've seen of the general process can be found in *The Story of Rope Making* (www.the-ropewalk.co.uk/ks2th2.pdf).

24. I'm ignoring some secondary matters such as the influence of the direction of the final twist on a rope's torsional behavior.

Chapter Ten

1. Or almost explicitly. Doing it right requires invoking calculus, in effect making those "little elements of mass" approach zero in size with a little mathematical sleight of hand.

2. Yes, you've every right to be puzzled. Mass may be mass, but moment of inertia depends on how something turns, that is, where the axis of rotation happens to be located for a particular motion.

3. A good general introduction is that of Edwards (1986). Di Bartolo (2010) gives a hairier physical analysis.

4. Good high-speed videos at this righting (search "cat righting reflex"): http://video.nationalgeographic.com/video/cats_domestic_ninelives?source=

relatedvideo; http://www.huffingtonpost.com/2012/08/24/why-cats-always-land
-on-their-feet-_n_1828748.html. The latter has a particularly clear explanation of
the physics involved.

5. Jusufi et al. (2008).

6. Or you can create several, whose vorticity adds up properly to exactly the
opposite of the one you care about.

7. I give an account that's more accessible than the one in typical fluid mechanics texts in Vogel (2013).

8. Smeaton (1759); the larger picture is definitively presented by Anderson
(1997).

9. A quite wonderful book about the celebrated cartoonist has recently
appeared, written by his granddaughter Jennifer George (with several other authors), *The Art of Rube Goldberg: (A) Inventive (B) Cartoon (C) Genius* (2013).

10. *The Repertory of Patent Inventions*, no. XVII, n.s. (May 1835): 277–90.

11. Production began in the '30s, by the Pennwood Electric Company. The
clocks were called Numechrons or, later, Numechron Tymeters. Pennwood became
Pennwood Numechron before being acquired by Spartus. eBay usually has plenty.

12. The *Wikipedia* article "Geneva drive" has a fine animation.

13. But one does appear as device 43 (of a total of 1,649!) in Hiscox
(1907), p. 25.

14. Good accounts, despite their age, are Disston and Sons (1916) and Mercer
(1929).

15. Good Scandinavian bow saws remain available, though—the current chief
brand is Bahco.

16. This from William Gurstelle, without further reference in the source I saw
(http://www.make-digital.com/make/vol25?pg=172#pg172). His book *Backyard
Ballistics* (2012) is loads of fun.

17. According to Joseph Needham (1965), p. 99, who credits Thomas Ewbank
(1842), p. 69, for this opinion.

18. Ewbank (1842), pp. 69–70. This is the twelfth edition, so his comments
may have originated earlier.

19. Tarkov (1986); Koeppel (2009).

Chapter Eleven

1. The operative variable is called the "Froude propulsion efficiency," after the
great nineteenth-century British naval engineer William Froude.

2. *Wikipedia* has a fine animation of its operation: http://en.wikipedia.org
/wiki/Newcomen_atmospheric_engine. Another can be found at http://www
.animatedengines.com/newcomen.html.

3. A real one, moved from the UK and restored, can be seen at the Henry
Ford Museum, in Dearborn, Michigan. It has a piston 28 inches in diameter.

4. The Rumseian Society has a nice set of diagrams at http://jamesrumsey.org
/how-the-steamboat-works/.

5. A good introduction is Liebowitz and Margolis (1995). "Lock-in" has been used both for inadvertent capture of a market and deliberate exclusion of competition by use of proprietary components, to my way of thinking quite different notions.

6. For allusion to other similar designs, see Cowan (1997).

7. The original few pages still read well. See Henry (1831).

8. These details from the *Wikipedia* article "Electric motor" and from M. Doppelbauer, of the Elektrotechnisches Institut of the Karlsruhe Institute of Technology, via http://www.eti.kit.edu/english/1376.php.

9. Isambard Kingdom Brunel has to be one of the most fascinating figures in the history of technology. A good biography is the one by Rolt (1957).

10. The *Wikipedia* articles "Pulsejet" and "Ramjet" provide good entry points for both pulse-jets and ramjets.

11. Once again, the *Wikipedia* article "Hydraulic ram" gives a good account of both history and operation. If you want to make one from commercially available parts, see http://www.clemson.edu/irrig/equip/ram.htm.

12. MacDonald (1945). It generated a movie (1947), a brief TV series, and, most recently, a restaurant chain—not, by the way, the one carrying the surname above.

13. As you go faster, you swing your legs farther, not more often. Eventually you reach the maximum practical swinging amplitude. Small animals can reach higher swinging rates with their shorter legs, but they still can't go as fast. The operative formula (a rough one, since animals vary in ways other than size) is that the maximum practical walking speed is 2.2 times the square root of hip-to-ground distance, where speed is in meters per second and distance is in meters—about 5 mph or 12-minute miles for a hip height of a meter. More in Vogel (2013).

14. The silk of the orb webs of spiders has especially low resilience, around 35 percent. That's handy for a stick-on-to-capture device, but it's probably tolerable only because the strands are thin and only loaded by single impulses. I elaborate on this elsewhere. Vogel (2013).

15. They may have been developed earlier in China, according to Needham (1954–2008), but then lost again. Works on the history of horology abound; my favorite is Landes (2000), notable for its general cultural perspective.

16. An excellent paper on clepsydrae is one by Mills (1982). Sand clocks work on the same principle and avoid the viscosity problem, but over time the passage of sand enlarges the orifice.

17. For a great deal more on the subject, see Vogel (1998).

Appendix

1. Two short papers give the real story on demonstration of the Coriolis-attributed bathtub vortex, Shapiro (1962) and Trefethen et al. (1965).

[References]

Agricola, G. 1556. *De Re Metallica*. Translated by H. C. Hoover and L. H. Hoover. 1912; reprint, New York: Dover, 1950.

Anderson, J. D. 1997. *A History of Aerodynamics*. Cambridge: Cambridge University Press.

Anthony, D. W. 2007. *The Horse, the Wheel, and Language*. Princeton, NJ: Princeton University Press.

Baker, I. O. 1903. *A Treatise on Roads and Pavements*. New York: John Wiley and Sons.

Ball, P. 2006. *The Devil's Doctor: Paracelsus and the World of Renaissance Magic and Science*. New York: Farrar, Straus and Giroux.

Barber, E. W. 1991. *Prehistoric Textiles*. Princeton, NJ: Princeton University Press.

Berg, H. C., and R. A. Anderson. 1973. Bacteria swim by rotating their flagellar filaments. *Nature* 245: 380–82.

Besson, J. 1569–78. *Théâtre des instrumens mathématiques et méchaniques*. Paris.

Bowden, F. P., and D. Tabor. 1973. *Friction: An Introduction to Tribology*. Garden City, NY: Doubleday Anchor.

Brackenbury, J. 1999. Fast locomotion in caterpillars. *J. Insect Physiol.* 45: 525–33.

Brearly, H. C. 1919. *Time Telling through the Ages*. New York: Doubleday, Page.

Bressel, E., and G. D. Heise. 2004. Effect of arm cranking on EMG, kinematic, and oxygen consumption responses. *J. Appl. Biomech.* 20: 129–43.

Bürger, W. 1984. The yo-yo: a toy flywheel. *Am. Sci.* 72: 137–42.

Carter, G. F. 1980. The metate: an early grain-grinding implement in the New World. In *Early Native Americans: Prehistoric Demography, Economy, and Technology*, edited by D. L. Browman, 21–39. The Hague: Mouton.

Caulfield, S. 1977. The beehive quern in Ireland. *J. Roy. Soc. Antiquaries Ireland* 107: 104–38.

Childe, V. G. 1954. Rotary motion. In *A History of Technology*, Vol. 1, edited by C. Singer, E. J. Holmyard, and A. R. Hill, 187–215. Oxford: Clarendon Press.

Choleva, M. 2012. The first wheelmade pottery at Lerna: wheel-thrown or wheel-fashioned? *Hesperia* 81: 343–81.

Coppa, A., L. Bondioli, A. Cucina, D. W. Frayer, C. Jarrige, J.-F. Jarrige, G. Quivron, M. Rossi, M. Vidale, and R. Macchiarelli. 2006. Early Neolithic tradition of dentistry. *Nature* 440: 755–56.

Cotterell, B., and J. Kamminga. 1990. *Mechanics of Pre-Industrial Technology*. Cambridge: Cambridge University Press.

Cowan, R. S. 1997. *A Social History of American Technology*. New York: Oxford University Press.

Crisman, K. J., and A. B. Cohn. 1998. *When Horses Walked on Water*. Washington, DC: Smithsonian Institution Press.

Crowfoot, G. M. 1954. Textiles, basketry, and mats. In *A History of Technology*, Vol. 1, edited by C. Singer, E. J. Holmyard, and A. R. Hill, 413–47. Oxford: Clarendon Press.

Cubitt, W. 1822. *Description of the Tread Mill Invented by Mr. William Cubitt of Ipswich for the Employment of Prisoners*. London: Longman et al.

Davidson, D. S. 1947. Fire-making in Australia. *Am. Anthropol.* 49: 426–37.

Dewald, E. T. 1932. *The Illustrations of the Utrecht Psalter*. Princeton, NJ: Princeton University Press.

Di Bartolo, S. 2010. Orientation change of a two-dimensional articulated figure of zero angular momentum. *Am. J. Phys.* 78: 733–37.

Diamond, J. M. 2005. *Guns, Germs, and Steel: The Fates of Human Societies*. New York: W. W. Norton.

Disston, H., and Sons. 1916. *The Saw in History*. Philadelphia: H. Disston and Sons.

Edwards, M. H. 1986. Zero angular momentum turns. *Am. J. Phys.* 54: 846-847.

Ekholm, G. F. 1946. Wheeled toys in Mexico. *Am. Antiquity* 11: 222–28.

Ensor, R. 2004. The loss of the *Stirling Castle* in the great storm of 1703 and the earliest archaeological evidence of a ship's steering wheel mechanism. *Mariner's Mirror* 90: 92–98.

Erickson, H. H. 2006. History of horse-whims, teamboats, treadwheels and treadmills. *Equine Veterinary Journal* 38: 83–87.

Ewbank, T. 1842. *A Descriptive and Historical Account of Hydraulic and Other Machines for Raising Water*. New York: Bangs, Platt, and Co.

Forbes, R. J. 1964. *Studies in Ancient Technology*. Vol. IV. Leiden: E. J. Brill.

Foster, G. M. 1959. The Coyotepec "molde" and some associated problems of the potter's wheel. *Southwest J. Anthropol.* 15: 53–63.

French, M. J. 1988. *Invention and Evolution: Design in Nature and Engineering*. Cambridge: Cambridge University Press.

Full, R., K. Earls, M. Wong, and R. Caldwell. 1993. Locomotion like a wheel. *Nature* 365: 495.

Gebeshuber, I. C., M. Drack, and M. Scherge. 2008. Tribology in biology. *Tribology* 2: 200–212.

George, J., et al. 2013. *The Art of Rube Goldberg*. New York: Harry N. Abrams.

Gilbert, K. R. 1954. Rope-making. In *A History of Technology*, Vol. 1, edited by C. Singer, E. J. Holmyard, and A. R. Hill, 451–55. Oxford: Clarendon Press.

Gilroy, G. C. 1853. *The History of Silk, Cotton, Linen, Wool, and Other Fibrous Substances*. New York: C. M. Saxton.

Gordon, J. E. 1978. *Structures, or Why Things Don't Fall Down*. London: Penguin.

Grant, U. S. 1885. *Personal Memoirs of U. S. Grant*. Vol. 1. New York: Library of America, 1990.

Gurstelle, W. 2012. *Backyard Ballistics*. 2nd ed. Chicago: Chicago Review Press.

Hacker, B. C. 1997. Greek catapults and catapult technology. In *Technology and the West*, edited by T. S. Reynolds and S. H. Cutcliffe, 49–65. Chicago: University of Chicago Press.

Hardie, J. 1824. *The History of the Treadmill*. New York: S. Marks.

Henry, J. 1831. On a reciprocating motion produced by magnetic attraction and repulsion. *Am. J. Science and Arts* 20: 340–43.

Henschel, J. R. 1990. Spiders wheel to escape. *South African J. Sci.* 86: 151–52.

Herlihy, D. 2004. *Bicycle: The History*. New Haven, CT: Yale University Press.

Hiscox, G. D. 1907. *Mechanical Movements: Powers and Devices*. 11th ed. New York: Norman W. Henley Publishing.

Hodges, H. 1970. *Technology in the Ancient World*. New York: Knopf.

Hough, W. 1890. Aboriginal fire-making. *Am. Anthropologist* 3: 359–72.

Hurt, R. D. 1982. *American Farm Tools: From Hand-Power to Steam-Power*. Manhattan, KS: Sunflower University Press.

James, W. 1906. Lecture II: What Pragmatism Means. In *William James: Writings 1902–1920*. New York: Library of America.

Johnson, W. 1992. Benjamin Robins, F. R. S. (1707–1751): New details of his life. *Notes Rec. R. Soc. Lond.* 46: 235–52.

Jusufi, A., D. I. Goldman, S. Revzen, and R. J. Full. 2008. Active tails enhance arboreal acrobatics in geckos. *Proc. Nat. Acad. Sci. USA* 105: 4215–19.

Kerr, R., and N. Wood. 2004. *Science and Civilization in China*. Vol. 5, Part 12 (continuation of Joseph Needham's treatise).

King, F. H. 1911. *Farmers of Forty Centuries*. Madison, WI: Mrs. F. H. King; repr., 1927, New York: Harcourt & Brace.

King, R. 2000. *Brunelleschi's Dome*. London: Chatto & Windus.

Kirby, W., and W. Spence. 1815–26. *An Introduction to Entomology; or, Elements of the natural history of insects: With plates.* 4 vols. London: Longman, Hurst, Rees, Orme, and Brown.

Koeppel, G. T. 2009. *Bond of Union: Building the Erie Canal and the American Empire*. New York: Da Capo Press.

Kraybill, N. 1978. Pre-agricultural tools for the preparation of foods in the Old World. In *Origins of Agriculture*, edited by C. A. Reed, 485–522. The Hague: Mouton.

LaBarbera, M. 1985. Mechanical properties of a North American aboriginal fishing line: the technology of a natural product. *Am. Anthropologist* 87: 625–36.

Landels, J. G. 2000. *Engineering in the Ancient World*. Berkeley: University of California Press.

Landes, D. S. 2000. *Revolution in Time*. Cambridge, MA: Harvard University Press.

Lawler, A. 2014. Sailing Sinbad's seas. *Science* 344: 1440–45.

Lay, M. G. 1992. *Ways of the World: A History of the World's Roads and of the Vehicles that Used Them*. New Brunswick, NJ: Rutgers University Press.

Lemon, H. 1968. The development of hand spinning wheels. *Antiquity* 1: 83–91.

Lessing, H.-E. 1998. The evidence against Leonardo's bicycle. In *Proceedings of the 8th International Cycle History Conference*, edited by N. Oddy and R. van der Plas, 49–56. San Francisco: Cycle Publishing.

Lewis, M. J. T. 1994. The origins of the wheelbarrow. *Technol. and Culture* 35: 453–75.

Liebowitz, S. T., and S. E. Margolis. 1995. Path dependence, lock-in, and history. *J. Law, Econ. Organiz.* 11: 205–26.

Lilienthal, O. 1889. *Birdflight as the Basis of Aviation*. [*Vogelflug als Grundlage die Flugkunst*], translated by A. W. Isenthal. Hummelstown, PA: Markowski International Publishers, 2001.

Lucas, A., and J. R. Harris. 1962. *Ancient Egyptian Materials and Industries*. London: Edward Arnold.

MacDonald, B. 1945. *The Egg and I*. Philadelphia: J. B. Lippincott.

MacIntosh, B. R., P. F. Gardiner, and A. J. McComas. 2005. *Skeletal Muscle: Form and Function*. 2nd ed. Champaign, IL: Human Kinetics.

Major, J. K. 1990. Water, wind and animal power. In *An Encyclopaedia of the History of Technology*, edited by I. McNeil, 229–71. London: Routledge.

Marchetti, C. 1979. A postmortem technology assessment of the spinning wheel: the last thousand years. *Technological Forecasting and Social Change* 13: 91–93.

Marsden, E. W. 1969. *Greek and Roman Artillery: Historical Development*. Oxford: Clarendon Press.

———. 1971. *Greek and Roman Artillery: Technical Treatises*. Oxford: Clarendon Press.

McGrail, S. 2001. *Boats of the World*. Oxford: Oxford University Press.

McGuire, J. D. 1896. A study of the primitive methods of drilling. *Rept. U.S. Nat'l. Mus. 1894*, 623–756.

McLaren, K. G., and D. Tabor. 1961. The frictional properties of lignum vitae. *Br. J. Appl. Phys.* 12: 118–20.

McShane, C., and J. A. Tarr. 2007. *The Horse in the City: Living Machines in the Nineteenth Century*. Baltimore: Johns Hopkins University Press.

Mercer, H. C. 1929. *Ancient Carpenters' Tools*. Doylestown, PA: Bucks County Historical Society. Repr., New York: Dover, 2000.

Mills, A. A. 1982. Newton's water clocks and the fluid mechanics of clepsydrae. *Notes Rec. R. Soc. Lond.* 37: 35–61.

Minetti, A. E. 1995. Optimum gradient of mountain paths. *J. Appl. Physiol.* 79 (5): 1698–703.

Moon, F. C. 2007. *The Machines of Leonardo Da Vinci and Franz Reuleaux*. Dordrecht: Springer.

Needham, J. 1954–2008. *Science and Civilisation in China*. Cambridge: Cambridge University Press.

———. 1965. *Science and Civilization in China*. Vol. 4, pt. 2. Cambridge: Cambridge University Press.

Nicastro, N. 2008. *Circumference: Eratosthenes and the Ancient Quest to Measure the Globe*. New York: St. Martin's Press.

Oberg, E., F. D. Jones, and H. L. Horton. 1984. *Machinery's Handbook*. 22nd ed. New York: Industrial Press.

Oliver, S. H., and D. H. Berkebile. 1974. *Wheels and Wheeling: The Smithsonian Cycle Collection*. Washington, DC: Smithsonian Institution Press.

Orton, C., P. Tyers, and A. Vince. 1993. *Pottery in Archaeology*. Cambridge: Cambridge University Press.

Oxford English Dictionary. 2nd ed. (continuous revision) New York: Oxford University Press.

Patterson, R. 1954. Spinning and weaving. In *A History of Technology*, Vol. 2, edited by C. Singer, E. J. Holmyard, and A. R. Hill, 191–220. Oxford: Clarendon Press.

Peng, D., ed. 1988. *Ships of China*. Beijing: Chinese Institute of Navigation.

Piggott, S. 1983. *Earliest Wheeled Transport*. Ithaca, NY: Cornell University Press.

Plymouth Cordage Company. 1916. *The Story of Rope*. North Plymouth, MA: Plymouth Cordage Company.

Prager, F. D., and G. Scaglia. 1972. *Mariano Taccola and His Book "De Ingeneis."* Cambridge, MA: MIT Press.

Ramelli, A. 1588. *The Various and Ingenious Machines of Agostino Ramelli*, translated by M. T. Gnudi and E. S. Ferguson. Baltimore, MD: Johns Hopkins University Press, 1976.

Repertory of Patent Inventions, No. XVII. New Series. 1835, May. London: Simpkin, Marshall, and Co., 277–90.

Richards, R. H. 1903. *Ore Dressing*. Vol. 1. New York: The Engineering and Mining Journal.

Ritti, T, K. Grewe, and P. Kessner. 2007. A relief of a water-powered stone sawmill on a sarcophagus at Hierapolis and its implications. *J. Roman Archaeol.* 20: 138–63.

Rolt, L. T. C. 1957. *Isambard Kingdom Brunel*. London: Longmans Green.

Roosevelt, T. 1914. *Through the Brazilian Wilderness*. New York: C. Scribner's Sons.

Rybczynski, W. 2000. *One Good Turn: A Natural History of the Screwdriver and the Screw*. New York: Simon and Schuster.

Ryder, M. L. 1968. The origin of spinning. *Antiquity* 1: 73–82.

Scott, L. 1954. Pottery. In *A History of Technology*. Vol. 1, edited by C. Singer et al., 376–412. London: Oxford University Press.

Shapiro, A. H. 1962. Bath-tub vortex. *Nature* 196: 1080–81.

Sleeswyk, A. W. 1981. Hand-cranking in Egyptian antiquity. In *History of Technology*. Vol. 6, edited by A. R. Hall and N. Smith, 23–37. London: Mansell Publishing.

Smeaton, J. 1759. An experimental enquiry concerning the natural powers of water and wind to turn mills, and other machines, depending on a circular motion. *Phil. Trans. R. Soc. Lond.* 51: 100–174.

Smith, C. B. 2006. *How the Great Pyramid was Built*. New York: HarperCollins.

Sobel, D. 1995. *Longitude: The True Story of a Lone Genius Who Solved the Greatest Scientific Problem of His Time*. New York: Penguin.

Steele, B. D. 1994. Muskets and pendulums: Benjamin Robins, Leonhard Euler, and the ballistics revolution. *Technology and Culture* 35: 348–82. Also in *Technology and the West*, edited by T. S. Reynolds and S. H. Cutcliffe, 145–79. Chicago: University of Chicago Press, 1997.

Strada, J. de. 1617–18. *Künstlicher Abriss allerand Wasser-Wind. Ross-und-Handmühlen*. Frankfurt.

Stuart, R. 1829. *Historical and Descriptive Anecdotes of Steam Engines*. London: Wightman and Cramp.

Takaoglu, T. 2005. Coskuntepe: an early neolithic quern production site in NW Turkey. *J. Field Archaeol.* 30: 419–33.

Tarkov, J. 1986. Engineering the Erie Canal. *Am. Heritage of Invention and Technology* 2 (1): 50–57.

Tarr, J. A. 1999. A note on the horse as an urban power source. *J. Urban Hist.* 25: 434–48.

Tarver, W. T. S. 1995. A reconstruction of an early medieval siege engine. *Tech. Cult.* 36: 136–67.

Taylor, C. R., S. I. Caldwell, and V. J. Rowntree. 1972. Running up and down hills: some consequences of size. *Science* 178: 1096–97.

Temple, R. 1998. *The Genius of China: 3,000 Years of Science, Discovery and Invention*. London: Prion Books.

Thompson, J., and A. Smith. 1877. *Street Life in London*. Repr., New York: B. Blom, 1969.

Thurston, R. H. 1894. *The Animal as a Machine and a Prime Motor, and the Laws of Energetics*. New York: John Wiley and Sons.

———. 1895. *A History of the Growth of the Steam-Engine*. 5th ed. London: Kegan Paul, Trench, Trubner & Co.

Tiedemann, E. J., and K. A. Jakes. 2006. An exploration of prehistoric spinning technology: spinning efficiency and technology transition. *Archaeometry* 48: 293–307.

Trefethen, L. M., R. W. Bilger, P. T. Fink, R. E. Luxton, and R. I. Tanner. 1965. The bath-tub vortex in the southern hemisphere. *Nature* 207: 1084–85.

Twain, M. 1869. *Innocents Abroad; or, The New Pilgrims' Progress*. New York: American Publishing Co. [Available digitized from Project Gutenberg, http://www .gutenberg.org/files/3176/3176-h/3176-h.htm.]

Valturio, R. 1472. *De Re Militari*. Verona, Italy: Johannes Nicolai de Verona.

Van Bueren, T. M. 2004. "The poor man's mill": a rich vernacular legacy. *J. Soc. Industrial Archeol.* 30: 5–23.

Veranzio, F. 1615. *Machinae Novae*. Munich: Heinz Moos Verlag, 1965.

Vitruvius (M. Vitruvius P. 1st c. BCE). *The Ten Books on Architecture*, translated by M. H. Morgan. Cambridge, MA: Harvard University Press, 1914.

Vogel, S. 1998. *Cats' Paws and Catapults: Mechanical Worlds of Nature and People*. New York: W. W. Norton.

———. 2001. *Prime Mover: A Natural History of Muscle*. New York: W. W. Norton.

———. 2002. A short history of muscle-powered machines. *Natural History* 111 (2): 84–91.

———. 2013. *Comparative Biomechanics*. 2nd ed. Princeton, NJ: Princeton University Press.

White, K. D. 1984. *Greek and Roman Technology*. Ithaca, NY: Cornell University Press.

White, L. T. 1962. *Medieval Technology and Social Change*. New York: Oxford University Press.

———. 1974. Technology assessment from the stance of a Medieval historian. *Am. Hist. Rev.* 79: 1–13.

Wiegand, J. H. 1963. Demonstrating the Weissenberg effect with gelatin. *J. Chem. Education* 40: 475–76.

Wilson, D. G. 2004. *Bicycling Science*. 3rd ed. Cambridge, MA: MIT Press.

Winchester, S. 2008. *The Man Who Loved China*. New York: HarperCollins.

Woodbury, R. S. 1961. *History of the Lathe to 1850*. Cambridge, MA: MIT Press.

———. 1972. *Studies in the History of Machine Tools*. Cambridge, MA: MIT Press.

Wu, X., C. Zhang, P. Goldberg, D. Cohen, Y. Pan, T. Arpin, and O. Bar-Yosef. 2012. Early pottery at 20,000 years ago in Xianrendong Cave, China. *Science* 336: 1696–700.

Zeitler, W. W. 2013. *The Glass Armonica: The Music and the Madness*. San Bernardino, CA: Musica Arcana.